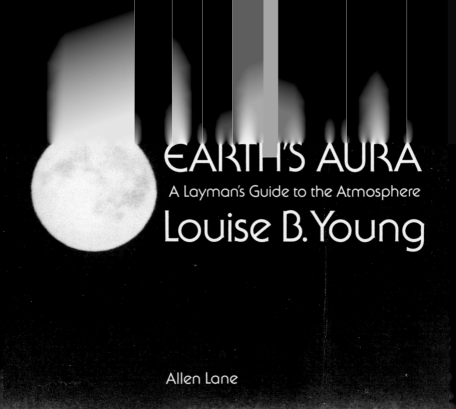

EARTH'S AURA

A Layman's Guide to the Atmosphere

Louise B. Young

Allen Lane

ALLEN LANE
Penguin Books Ltd
536 King's Road
London SW10 0UH

First published in the U.S.A. by Alfred A. Knopf 1977
First published in Great Britain simultaneously by Allen Lane
and Penguin Books 1980
Copyright © Louise B. Young, 1977

ISBN 0 7139 1165 4

Set in Monophoto Ehrhardt
Printed in Great Britain by
Hazell Watson & Viney Ltd, Aylesbury, Bucks

The acknowledgements on page 305
constitute an extension of this copyright page

To my husband,
who has never failed to encourage and help me,
contributing generously of his own scientific knowledge
and giving me that most precious of all gifts –
leisure time in which to pursue my research and writing.

LIST OF PLATES

LIST OF DIAGRAMS

All diagrams are reproduced by courtesy of Brian Marsh.

1. In this photograph taken from the Gemini VII spacecraft, the atmosphere appears as a thin luminous band of blue above the earth's surface.

INTRODUCTION

THE VIEW FROM SPACE

The earth is now passing by my window. It's about as big as the end of my thumb.

– Astronaut in Apollo 8[1]*

A traveller coming in from outer space to visit the earth would be struck first by one of the most unusual features of this little planet – the soft blue glow that makes it look like a luminous raindrop set against the blackness of space. The light from most heavenly bodies is diamond bright or, where they shine by reflected light as in the case of our moon, the light is less intense but still sharply defined. Only the planets that hold an appreciable atmosphere return the light that they reflect, altered – softened and coloured by its passage through the various layers of vapour. Venus is pale yellow, Jupiter a glowing gold and red. Saturn is set in iridescent rings of ice, and the earth is wrapped in a swirl of ethereal blue and white.

This is our sky – our blue heavens – seen inside out. It is the earth's aura. Not just a collection of gases and vapour, as we sometimes think of the atmosphere, it is a complex medium suffused with colour and energy and possessing qualities distinctive of the body from which it emanates. [See Plate 1.]

As the traveller from outer space comes closer to the earth,.he can see the profile of the aura, outlined between the curving surface of the planet and the black void beyond. He discovers how slender is this fragile layer of atmosphere that forms the nourishing and protecting medium for all earth's living things. It shelters them from the fierce heat and cold of space. It filters out the damaging rays of sunlight and burns up several million billion meteors each day to harmless cinders before they reach the earth's surface. From the oceans and lakes it draws up the essential life ingredient, water, distills it and purifies it and distributes it again across the planet's surface. Mobile and impressionable, it is a medium constantly on the move, creating the infinite variety of conditions on earth, from winds that can uproot trees and tear clothes from a man's

*Factual information and quotations are referenced in the Notes section beginning on page 281.

back to the gentlest of summer breezes, from paralysing cold that would freeze an unprotected man in less than ten minutes to heat that can turn pools of water into steam.

The earth's aura makes our seasons and our sunsets. It adds glowing colour where there would otherwise be blinding white light, too harsh to look upon. It holds and scatters the light long after the sun has set, making that lovely transition time, twilight. Unlike the sharp-edged light and darkness of space, we have hours that are neither day nor night – dusk and dawn. And the edges of things like the edges of time are softened by the air that envelops them, making their shapes and colours flow one into another.

Not only does the atmosphere alter the conditions on earth and make life possible, it is also formed and changed by them. Air is the hot breath of the earth itself poured forth from fiery volcanoes. It is the cumulative breath of all the living plants and animals on earth. Even the tiny bacteria who live in darkness under the earth's surface make their contribution to the composition of the earth's atmosphere. Man also, with his furnaces and his agriculture, changes earth's aura just as surely as the important conditions of his life are created by it.

And yet we who are born on the earth's surface and spend all our hours deeply immersed in the atmosphere take it so much for granted that we are almost unconscious of its existence. We use the words 'thin air' synonymously with 'emptiness'. And we speak of the 'limitless blue sky' as though it were space. We hardly notice the impact of the air on our senses. We are conscious only of the changes caused by its movement and the variations in its composition. We feel the wind and see the clouds and hear the waves breaking on the shore. Actually, the image of everything we see is affected by the air which diffuses the light. All the smells of the earth – the fragrance of flowers and the acrid smoke from industrial stacks – are nothing but chemicals carried in the atmosphere. And were it not for the air, there would be no sound at all, for sound is caused by waves of compression in the air. Although we think we cannot feel it, the atmosphere presses down on every square inch of our bodies with the weight of almost 15 pounds. This medium in which we live is so all-pervasive that it is the norm and, therefore, the nothingness.

A fish at the bottom of the sea must have the same impression of water. It does not really notice the water itself as much as the changes in it – the eddies that stir the sand on the ocean floor, the backwash of the waves that break over the rocks and swirl past the coral reef, the current of warm water that rushes northwards from the tropical seas. The fish is not conscious of the way the water buoys up its body, counteracting the force of gravity (which indeed it has never experienced). So water would be nothingness to a fish just as air is to a man or a bird or a flower. Only a flying fish that breaks free of its medium, sailing in a long arc over the surface, can look down at the water's shining ruffled surface, see the deep indigo colour, and feel the thick slipperiness of the liquid flow about its body again when it dives back in.

So man had to break free of his own medium and see it from outside before he could truly grasp its quality and significance.

'What beauty,' exclaimed Yuri Gagarin, the first man to look back on the planet from space. 'I saw clouds and their light shadows on the distant dear earth . . . The water surface looked like darkish, slightly gleaming spots . . . When I watched the horizon, I saw the abrupt, contrasting transition from the earth's light-colored surface to the absolutely black sky. I enjoyed the rich color spectrum of the earth. It is surrounded by a light blue aureole that gradually darkens, becomes turquoise, dark blue, violet, and finally coal black.'[2]

The Space Age has given us this new perspective. In a few short decades we have gained much knowledge. Some mysteries have been solved and some have only deepened. The picture that has emerged is rich in variety and complexity. We have discovered that the atmosphere is moved by dynamic processes that were never suspected before man went into space. We have photographed the ghostly green glow of the ionosphere shining by its own light at night and have measured the current of solar wind that girdles the globe. Hundreds of miles above the veil of clouds where the atmosphere appears to fade away into infinity are regions where the earth's aura is still charged with power and energy, responding, in ways still just dimly understood, to the unexplained forces that flow in upon it from the farthest reaches of the universe.

CHAPTER ONE

MAN PROBES THE ATMOSPHERE

*For we are dwelling in a hollow of the earth, and fancy that we are on the surface . . .
But the fact is, that owing to our feebleness and sluggishness we are prevented from
reaching the surface of the air.*

– Plato[1]

One night in 1818, or thereabouts, in Lancaster, Pennsylvania, a young
boy stood at his open bedroom window in his nightdress looking up at
the storm-torn sky. Clouds scudded across the waning moon and now and
then they parted to show a view of stars. He could just make out the faint
red glow of the planet Mars and the hazy arc of the Milky Way. As he
watched, a meteor suddenly marked a silvery path across the patch of sky
and disappeared behind a giant thunderhead where lightning flickered
like a guttering candle.

John Wise was thinking about his kitten, Nellie. Was she up there still
in that flickering cloud, tossed and twisted by wind, drenched by rain,
dazzled by the lightning flash? Or was she sailing smoothly in the clear
open spaces between the clouds, in that mysterious and wonderful sky
high above the earth where the lights of the towns must look milky and
far away like clusters of stars!

That afternoon John had taken out his new kite, the one he had been
working on for several weeks and which was the largest that he had ever
built. He prepared it for flight, attaching the long knotted tail. Then he
wrapped up his kitten and tied her securely to the kite. He was curious to
see whether the kite would be strong enough to lift Nellie into the sky.

The afternoon was a little threatening and a gusty wind picked the kite
up from the ground. In a flash it rose far above the trees and sailed high.
Nellie had broken loose from the bonds of the earth's gravity and taken
off on a remarkable adventure. Soon, John thought, he would have to reel
the kite back in. That cloud looming to the west was growing darker
every minute and he could hear the muffled roll of distant thunder.
Suddenly the kite wrenched violently at John's hand and broke loose. The

dark speck dipped and then spiralled straight upwards. Nellie sailed off into the thunderstorm.

Two days later Nellie returned home, limping slightly and looking a little bedraggled but otherwise none the worse for wear. How tantalizing it was that she was unable to tell anything about those strange hours when the earth had receded beneath her and she had seen the inside of a thunderhead! One simple fact John had learned by his experiment – an animal can go high above the earth's surface and survive.

Although John did not know it at the time, animals and even men had been lifted to altitudes of several thousand feet in lighter-than-air balloons. But perhaps none of them had flown as high as Nellie. The updraft in a thunderstorm could have lifted the kite to an altitude of ten or fifteen thousand feet.

Later John started to build a kite which he hoped would be large enough to carry his own weight, but before the project was completed he happened to see a newspaper article describing a balloon flight in Italy. He quickly revised his plans and started work on a lighter-than-air balloon. During his lifetime Wise made nearly five hundred ascensions; he survived one occasion when his balloon caught fire and he made several trips of his own into thunderstorms. Finally, at the age of seventy-two, he was lost in a balloon which was last seen hovering over the southern end of Lake Michigan. The body of his companion was found washed up on the Indiana shore but the remains of the balloon and Wise's body were never recovered.[2]

John Wise had been fond of explaining to people that on the earth's surface they carried an ocean of air on their shoulders. 'Come up with me two or three miles,' he would urge then, 'the lightened load will round out your eyeballs. You will be able to read without spectacles.'[3]

At the beginning of the nineteenth century it was well understood that the air far above the earth was thinner and lighter than the air at the earth's surface. Torricelli and Pascal had discovered this important fact a hundred and fifty years earlier. Up until that time – for almost two million years – man had lived immersed in a body of air that weighs down upon the earth with a force of five thousand million million tons and had been almost unconscious of its existence. The process of evolution had so perfectly adapted the human body to exist and function in the presence of this pressure that actually a man could not live without it. Remove the atmosphere and a man's body would swell, his blood would begin to boil, the vessels would rupture, and death would soon follow.

It was indeed the effect of the absence of air that was noticed first. When people tried to draw liquids such as wine from a large container, they found that the wine would not flow out unless two holes were made, one in the top and one in the bottom of the container. 'Nature abhors a vacuum,' explained Aristotle. If you let wine out, you must let air in, because a vacuum is impossible. And that explanation was accepted for many centuries.

In 1638 Galileo made the observation that a suction pump such as those that were used to bring water up from deep wells could not raise water more than 34 feet. But in spite of his brilliant grasp of many other physical phenomena, Galileo did not recognize the significance of this observation. It was his pupil Torricelli who drew the right conclusion. The fact that water is raised 34 feet and no more must be a measure of the weight of the atmosphere, Torricelli said, and he confirmed this theory with an experiment. Mercury is 14 times heavier than water, so if a column of water 34 feet high is held up by the pressure of the atmosphere, then a column of mercury 1/14th that high should also be held up by it. Torricelli took a glass tube three feet long and filled it with mercury. He put his finger over the open end, inverted the tube, and submerged the end in a bowl of mercury. A little of the mercury flowed out of the tube until the height of the column was about 30 inches. Then no more mercury flowed into the bowl. The empty space at the top of the glass tube was a vacuum. Torricelli observed and measured the column for several days and he found that there were small variations in the height. It was not until several decades later that the relationship between these small variations and weather was observed and explained. But even though Torricelli did not understand all of these implications, he had demonstrated the existence of air pressure and invented the prototype of the altimeter and the barometer.[4]

News of Torricelli's experiment reached the French philosopher and scientist Blaise Pascal, and he repeated the experiment in Rouen. Pascal was especially interested in the theory of water pressure and had written a book explaining the laws of hydrostatics. Torricelli's experiment seemed to verify the theory that we live in a sea of air which exerts pressure on everything that exists in it, just as water exerts pressure. If that is so, reasoned Pascal, similar laws should apply. For example, as we rise higher above the earth the pressure should diminish, just as the pressure is less on an object slightly below the surface of the ocean than it is at the bottom. The two media are not exactly alike, of course; water is much heavier than air and air is compressible. Pascal thought of the atmosphere in terms of an analogy to a mass of wool, as he wrote in 1648:

> If there were collected a great bulk of wool, say twenty or thirty fathoms high, this mass would be compressed by its own weight; the bottom layers would be far more compressed than the middle or top layers, because they are pressed by a greater quantity of wool. Similarly the mass of the air, which is a compressible and heavy body like wool, is compressed by its own weight, and the air at the bottom, in the lowlands, is far more compressed than the higher layers on the mountaintops, because it bears a greater load of air.
>
> In the case of that bulk of wool, if a handful of it were taken from the bottom layer, compressed as it is, and lifted, in the same state of compression, to the middle of the mass, it would expand of its own accord; for it would then be nearer the top and subjected there to the pressure of a smaller quantity of wool. Similarly if a body of air, as found here below in its natural state of compression, were by some device transferred to a mountaintop, it would necessarily expand and come to the condition of the air around it on the mountain; for then it would bear a lesser weight of air than it did below.

As he thought about this analogy an exciting idea occurred to Pascal. Torricelli's column of mercury could be used to measure the pressure of the atmosphere at two different elevations – for instance, in the lowlands and on a mountaintop. And if there were a measurable difference in the height of the mercury at the two places, then the theory of the ocean of air would have been confirmed.

> . . . I conceived at that very time the experiment here described, which I hoped would yield definite knowledge as a ground for my opinion. I have called it the Great Experiment on the Equilibrium of Fluids, because it is the most conclusive of all that can be made on this subject, inasmuch as it shows the equilibrium of air and quicksilver, which are,

respectively, the lightest and the heaviest of all the fluids known in nature.

But since it was impossible to carry out this experiment here, in the city of Paris, and because there are very few places in France that are suitable for this purpose and the town of Clermont in Auvergne is one of the most convenient of these, I requested my brother-in-law, Monsieur Périer, counselor in the Court of Aids in Auvergne, to be so kind as to conduct it there.[5]

After making this request Pascal waited impatiently while months went by with no word about the experiment. Finally he received the following letter from Monsieur Périer:

Monsieur,   22 September 1648

At last I have carried out the experiment you have so long wished for. I would have given you this satisfaction before now, but have been prevented both by the duties I have had to perform in Bourbonnais, and by the fact that ever since my return the Puy-de-Dôme, where the experiment is to be made, has been so wrapped in snow and fog that even in this season, which here is the finest of the year, there was hardly a day when one could see its summit, which is usually in the clouds and sometimes above them even while the weather is clear in the plains. I was unable to adjust my own convenience to a favorable state of the weather before the nineteenth of this month. But my good fortune in performing the experiment on that day has amply repaid me for the slight vexation caused by so many unavoidable delays.

I send you herewith a complete and faithful account of it, in which you will find evidence of the painstaking care I bestowed upon the undertaking, which I thought it proper to carry out in the presence of a few men who are as learned as they are irreproachably honest, so that the sincerity of their testimony should leave no doubt as to the certainty of the experiment.

THE ACCOUNT OF THE EXPERIMENT
SUBMITTED BY MONSIEUR PERIER

The weather on Saturday last, the nineteenth of this month, was very unsettled. At about five o'clock in the morning, however, it seemed sufficiently clear; and since the summit of the Puy-de-Dôme was then visible, I decided to go there to make the attempt. To that end I notified several people of standing in this town of Clermont, who had asked me to let them know when I would make the ascent. Of this company some were clerics, others laymen. Among the clerics was the Very Rev. Father Bannier, one of the Minim Fathers of this city, who has on several occasions been 'corrector' (that is, father superior), and Monsieur Mosnier, canon of the Cathedral Church of this city; among the laymen were Messieurs la Ville and Begon, councilors to the Court of Aids, and Monsieur la Porte, a doctor of medicine, practicing here. All these men are very able, not only in the practice of their professions, but also in every field of intellectual interest. It was a delight to have them with me on this fine work.

On that day, therefore, at eight o'clock in the morning, we started off all together for the garden of the Minim Fathers, which is almost the lowest spot in the town, and there began the experiment in this manner.

First, I poured into a vessel six pounds of quicksilver which I had rectified during the three days preceding: and having taken glass tubes of the same size, each four feet long and hermetically sealed at one end but open at the other, I placed them in the same vessel and carried out with each of them the usual vacuum experiment. Then, having set them up side by side without lifting them out of vessel, I found that the quicksilver left in each of them stood at the same level, which was twenty-six inches and three and a half lines above the surface of the quicksilver in the vessel. I repeated this experiment twice at this same spot, in the same tubes, with the same quicksilver, and in the same vessel; and found in each case that the quicksilver in the two tubes stood at the same horizontal level, and at the same height as in the first trial.

That done, I fixed one of the tubes permanently in its vessel for continuous experiment. I marked on the glass the height of the quicksilver, and leaving that tube where it stood, I requested the Rev. Father Chastin, one of the brothers of the house, a man as pious as he is capable, and one who reasons very well upon these matters, to be so good as to observe from time to time all day any changes that might occur. With the other tube and a portion of the same quicksilver, I then proceeded with all these gentlemen to the top of the Puy-de-Dôme, some five hundred fathoms above the convent. There, after I had made the same experiments in the same way that I had made them at the Minims', we found that there remained in the tube a height of only twenty-three inches and two lines of quicksilver; whereas in the same tube, at the Minims', we had found a height of twenty-six inches and three and a half lines. Thus between the heights of the quicksilver in the two experiments there proved to be a difference of three inches one line and a half. We were so carried away with wonder and delight, and our surprise was so great, that we wished, for our own satisfaction, to repeat the experiment. So I carried it out with the greatest care five times more at different points on the summit of the mountain, once in the shelter of the little chapel that stands there, once in the open, once shielded from the wind, once in the wind, once in fine weather, once in the rain and fog which visited us occasionally. Each time I most carefully rid the tube of air; and in all these experiments we invariably found the same height of quicksilver. This was twenty-three inches and two lines, which yields the same discrepancy of three inches, one line and a half in comparison with the twenty-six inches, three lines and a half which had been found at the Minims'. This satisfied us fully.

Later, on the way down at a spot called Lafon de l'Arbre, far above the Minims' but much farther below the top of the mountain, I repeated the same experiment, still with the same tube, the same quicksilver, and the same vessel, and there found that the height of the quicksilver left in the tube was twenty-five inches. I repeated it a second time at the same spot; and Monsieur Mosnier, one of those previously mentioned, having the curiosity to perform it himself, then did so again, at the same spot. All these experiments yielded the same height of twenty-five inches, which is one inch, three lines and a half less than that which we had found at the Minims', and one inch and ten lines more than we had just found at the top of the Puy-de-Dôme. It increased our satisfaction not a little to observe in this way that the height of the quicksilver diminished with the altitude of the site.

On my return to the Minims' I found that the [quicksilver in the] vessel I had left there in continuous operation was at the same height at which I had left it, that is, at twenty-six inches, three lines and a half; and the Rev. Father Chastin, who had remained there as observer, reported to us that no change had occurred during the whole day, although the weather had been very unsettled, now clear and still, now rainy, now very foggy, and now windy.

Here I repeated the experiment with the tube I had carried to the Puy-de-Dôme, but in the vessel in which the tube used for the continuous experiment was standing. I found that the quicksilver was at the same level in both tubes and exactly at the height of twenty-six inches, three lines and a half, at which it had stood that morning in this same tube, and as it had stood all day in the tube used for the continuous experiment.

I repeated it again a last time, not only in the same tube I had used on the Puy-de-Dôme, but also with the same quicksilver and in the same vessel that I had carried up the mountain: and again I found the quicksilver at the same height of twenty-six inches, three lines and a half which I had observed in the morning, and thus finally verified the certainty of our results . . .

If you find any obscurities in this recital I shall be able in a few days to clear them up in conversation with you, since I am about to take a little trip to Paris, when I shall assure you that I am, Monsieur,

Your very humble and very affectionate servant,

Périer[6]

This experiment dramatically supported Pascal's theory and afforded him 'great satisfaction'. In the meantime, M. Périer and his 'irreproachably honest' friends had had the unusual experience of conducting a successful and significant scientific experiment.

Pascal had taken advantage of the easiest way to probe the higher atmosphere – by climbing a mountain. But there is an obvious limitation to this method. The highest mountains on earth are about 29,000 feet and these forbidding towers of rock and ice are not suitable sites for scientific experimentation. The next step into space was taken in the hot-air balloon.

It was well known that hot air rises, a fact that can be observed by watching the smoke from a fire or feeling the hot updraft from a candle. The reason for this phenomenon can be easily understood by remembering that air (or anything in the gaseous state) contains countless billions of molecules all free to move independently and all moving in a random way at enormous speeds. They constantly collide with each other and anything else that happens to be in the way. The average velocity of their motion is what we measure as temperature. Each molecule is subject to the attraction of gravity but resists this force by virtue of its speed. The faster the molecules are moving, the more they tend to spread out, and if they are in a flexible container like a balloon, the greater the speed of the molecules, the more inflated the balloon becomes. The hot air in the balloon is lighter than the air it displaces and, therefore, it rises.

The buoyancy provided by hot air was first exploited in 1783 by two French brothers, Étienne and Joseph Montgolfier, who manufactured paper in the town of Annonay near Lyons. They built an enormous bag out of linen, lined it with paper, and filled it with hot air by burning a straw fire under it – a hazardous undertaking which miraculously did not end in disaster. On 21 November the Montgolfier balloon carried two men in a demonstration flight over the city of Paris. Throughout the flight the crew was frantically stoking the fire with straw. At one point the envelope did catch fire but the crew managed to extinguish the blaze. After travelling eight miles and reaching a height of 3,000 feet, they landed safely on the outskirts of the city.

The next balloons that were built were filled with hydrogen. Hydrogen is the lightest element; its atom consists of just one particle in the nucleus, a proton, and one electron. In the gaseous state atoms of hydrogen travel in pairs, making up the hydrogen molecule. A given volume of hydrogen is about sixteen times lighter than air, so the problem of providing a continuous stream of hot air to lift a balloon could be eliminated. Of course, hydrogen can be dangerous if not properly handled; a spark and a whiff of air is enough to touch off an explosion.

Hydrogen balloons carried men in little open wicker baskets to astonishing altitudes. They also carried dogs and sheep, ducks and pigeons, to demonstrate the 'wholesomeness' of this new environment. In almost every flight some modification of Torricelli's mercury tube went aloft to measure the pressure of the atmosphere at different altitudes. Just as Pascal had predicted, the height of the column of mercury declined regularly as the balloon rose and this fact was used to measure altitude. Not only was there less air above the balloon but the air near the earth was more compressed so that there were more molecules per cubic inch. Exactly how thin the air becomes at different heights was not known until these flights were made. Nor was it understood that a person could not travel very far above the earth before the reduced air pressure caused the body to swell and bubbles to form in the bloodstream.

Thermometers were standard equipment on balloon trips and temperatures were found to drop off very rapidly as higher altitudes were attained.

Some attempts were made to sample the air at different altitudes. Bottles were carried aloft, filled, and sealed at various heights above the ground. These rather crude experiments indicated that there was no great change in the composition of the air at altitudes up to about five miles. At these great heights air still contained many bacteria, grains of pollen, particles of soot, salt crystals, and even grains of sand. It still contained water vapour and nitrogen and the essential ingredient, oxygen. Oxygen (which was sometimes called 'fire air' or 'life air') had been discovered in 1774. It is one of the most reactive of all elements, having a strong tendency to combine with other substances, sometimes so violently that fire and explosions occur. In fact, oxygen is a necessary ingredient of any burning process. It was also known that oxygen is absolutely essential to most forms of life. Deprive an animal of oxygen and it dies.

An atom of oxygen is much heavier than hydrogen – sixteen times as heavy – and in the normal gaseous state oxygen travels in pairs just as hydrogen does. Since the way in which the gases are distributed above the earth is the result of two conflicting processes – the tendency to fall to the ground because of their weight and the tendency to keep moving, which depends upon their temperature – scientists recognized that the heavier gases like oxygen might thin out more rapidly with increasing altitude than the lighter constituents of air. Hydrogen gas is present only in trace amounts in the atmosphere. Nitrogen, which is by far the most common element in the air (comprising 78 per cent), is slightly lighter than oxygen. At what point is the oxygen too thin to support life? Scientists were anxious to know the answer to this question.

Several of the early aeronauts reported strange and alarming symptoms when their balloons attained high altitudes. An Italian, Count Francesco Zambeccari, flying with two friends in a hot-air balloon over the Adriatic in 1804, reported that the balloon suddenly rose:

all at once, but with such rapidity and to such a prodigious elevation that we had difficulty in hearing each other – even when shouting at the top of our voices. I was ill and vomited; Grassetti was bleeding at the nose. We were both breathing short and hard, and felt oppression on the chest. Because the balloon rose so suddenly out of the water and bore us with such swiftness to those high regions, the cold seized us suddenly, and we found ourselves covered with a layer of ice.[7]

The Italians were unable to report the altitude at which these effects had occurred. They had thrown all their instruments, including their altimeter, overboard earlier in the flight when the balloon threatened to descend into the stormy Adriatic Sea. The symptoms that they suffered could have signalled insufficient oxygen but at the time the favoured explanation was that these aeronauts had ascended too rapidly to become acclimatized to the high altitude.[8]

In 1862 the British Association for the Advancement of Science decided to send one of their colleagues up to investigate these problems and make scientific measurements. James Glaisher, a scientist of some reputation who had founded the British Meteorological Society and helped to set up the Aeronautical Society of Great Britain, volunteered for the flight. He was a dignified gentleman, fifty-three years old, with a bald head and mutton-chop whiskers. Glaisher was joined in the experiment by Henry Coxwell, who, at forty-three, had become known as the best aeronaut in England.

Coxwell built the balloon. It was fifty-three feet in diameter and designed to be filled with a gas made from the distillation of coal. This gas is lighter than air and is composed chiefly of hydrogen and methane (a compound of carbon and hydrogen). Like most aerostats, it was fitted with a valve which was controlled from the basket by a cord and could be opened to allow the gas to escape when the crew wanted to descend. The expedition carried a number of scientific instruments for measuring altitude, humidity, temperature, the earth's magnetic and electric field, and the solar spectrum. Sandbags were carried for ballast and a crate of pigeons for testing the effect of high altitude on other living things.

On the morning of 5 September 1862, Coxwell and Glaisher readied their balloon for flight. The weather was threatening but they dared not delay their departure again. It had already been postponed three times and the representatives of the Balloon Committee of the British Association for the Advancement of Science were growing impatient. Black clouds loomed overhead and a very strong ground wind was blowing when Glaisher and Coxwell climbed into the basket with their load of fragile instruments and the crate of pigeons. No sooner had they cut loose than a driving rain began and the balloon was bounced roughly across the launching field. Then suddenly it was seized by a current of air and carried steeply upwards into the clouds. The two men were drenched by torrents of rain, buffeted by gusts of wind. Emerging briefly into a brighter sky, they saw a gigantic yellow thunderhead just

2. *The dramatic cloud formations of a developing thunderhead.*

overhead. In another instant they were sucked up into it. Their unwieldy craft was spun violently upwards in the vortex of a giant maelstrom. Thunder and lightning crashed about them as they crouched against the sides of their basket, trying to shield the bundle of delicate scientific instruments with their bodies.

Then all at once the sky lightened; the sun shone through, pale and milky, and they emerged on one side of the storm. They could see it still lowering high behind them, but ahead they looked out on a dazzling world of brilliant blue sky and mountainous white clouds that looked soft and inviting as enormous piles of down. They were at 10,000 feet, Glaisher announced as he checked his instruments. Miraculously, none of them had been lost or damaged in the storm. The temperature hovered around the freezing point and the men buttoned up their coats to protect themselves from the cold wind. Coxwell emptied three bags of sand over the rim of the basket and the balloon rose swiftly in response.

Glaisher began taking his measurements as calmly and methodically as if he had been snug and warm in his laboratory. When they reached 15,000 feet the water in the wet-bulb thermometer began to freeze, making the instrument useless. But the dry-bulb thermometer measured 18°F (−8°C). Glaisher released one of the pigeons. It extended its wings, then fluttered awkwardly and somewhat out of control towards the earth.

At an altitude of 20,000 feet a second pigeon was released. It fell a considerable distance before it began to fly very slowly downwards in a spiral pattern.

At 25,000 feet the temperature had dropped far below zero. Now both Coxwell and Glaisher had difficulty performing the simplest tasks. It was hard to focus on the instruments and to record the numbers; it was an almost impossible task to pour out another bag of sand. Coxwell opened the pigeons' cage to let the last two birds out, but they huddled together at the back of the cage. Finally he shoved them out. One fell straight down, never opening its wings at all. The other fluttered frantically, managing at last to fly up towards the top of the balloon, but it was never seen again.

The balloon was still rising. It had reached 28,000 feet when Glaisher noticed that the cord controlling the release valve was out of reach, entangled in the rigging. There was no way to let the gas out. They could not bring the balloon back down to earth, at least not until the sun went down, permitting the gas to cool and shrink. But sunset was many hours away.

Both Glaisher and Coxwell stood up, studying the problem. The basket was hung from a large iron hoop, and the shroud ropes connected the hoop to the balloon. By climbing onto this hoop, Coxwell thought, he could just reach the tangled valve cord. He must try, although every step, every movement was now incredibly difficult. He hoisted himself first onto the edge of the basket, balancing there 28,000 feet above the earth. Then, with the help of the shroud ropes, he pulled himself slowly upward until he was able to hook one leg over the iron hoop. By careful manoeuvring he pulled his other leg across and rested, grasping the hoop tightly to steady himself. It was bitterly cold; the neck of the balloon was white with frost. Coxwell saw that the valve cord was now within reach. He tried to unclench his hands to grasp it but he could not free them. They were frozen fast to the iron hoop.

Watching Coxwell, Glaisher did not understand what had gone wrong. He had trouble seeing clearly; there seemed to be a fog before his eyes. 'I dimly saw Mr Coxwell,' he recalled later,

> and endeavoured to speak, but could not. In an instant intense darkness overcame me, so that the optic nerve lost power suddenly, but I was still conscious, with as active a brain as at the present moment whilst writing this. I thought I had been seized with asphyxia and believed I should experience nothing more, as death would come unless we speedily descended: other thoughts were entering my mind when suddenly I became unconscious as on going to sleep.[9]

Coxwell called out to Glaisher but got no answer. He tried desperately to drag his hands loose from the hoop but could not free them. They had turned dark – almost black with cold. Coxwell felt no pain, but an overwhelming sleepiness. Just over his head the valve line swung in a slow circle. With an enormous effort of will, he lunged and tried to catch it in his teeth. Twice it eluded him. Then, finally, he caught the line in his mouth. He wrenched it so hard that he broke a tooth but the valve did not open. Still holding the cord in his teeth, Coxwell used his tongue to push it back towards the side of his mouth where he could grasp it with his molars. Then he pulled again. This time the line cut deeply into his cheek. His mouth filled with blood and suddenly nausea overcame him. He vomited, losing the valve line and all the contents of his stomach into space. Finally the wave of sickness passed. He seized the valve cord again, clamping his teeth down hard on it, and gathering his last ounce of strength, he threw himself backwards. The valve broke open suddenly, his hands tore free. Somersaulting through space, he landed in the basket.

The line was free now, hanging so low that Coxwell could reach it easily. He caught it again in his mouth and, pulling gently, he heard the soft whistle of escaping gas. The balloon slowly began its descent.

It was many minutes before Glaisher, paralysed by lack of oxygen, began to recover his sight. He could hear Coxwell urging him, 'Do try; now do,' but he could not speak or move. Then, very gradually, as the descent continued, all his faculties returned.

Both aeronauts recovered from this extraordinary trip to the very limits of space in which unprotected human beings can survive. Before Glaisher had passed out he had taken a reading of 30,000 feet but had been unable to record this number. Coxwell thought that he had seen the mercury in the barometer stand at seven inches, which would have been 35,000 feet, but his condition at the time threw doubt upon that reading.[10] Either altitude was a remarkable achievement without the aid of oxygen equipment. Today pilots are warned not to fly above 10,000 feet for long in an unpressurized plane without oxygen. A volume of air at 18,000 feet contains only about half as much oxygen as at sea level. At 55,000 feet an unprotected pilot becomes unconscious in a few seconds; longer exposure can lead to brain damage and death. After Glaisher and Coxwell's experience aeronauts did not venture into the upper reaches of the atmosphere without supplies of oxygen.

The effects of low pressure were still not well understood and it was several decades before pressurized gondolas or special space suits were invented. The first sealed and pressurized capsule capable of protecting the aeronaut from the effects of low pressure and extreme cold was designed by a Swiss physicist, Auguste Piccard. In this spherical aluminium gondola, Auguste and his twin brother, Jean, were able to attain an altitude of 51,793 feet in 1931. In the two succeeding decades this and other inventions made it possible for men to probe the upper atmosphere up to heights of 100,000 feet.

Long before that time a much safer lifting element than hydrogen had been discovered. In the last years of the nineteenth century a new element was identified in the sun. It was named *helium*, from *helios*, the Greek word for the sun, and was eventually found to comprise a very small percentage of the earth's atmosphere (1 part in 186,000). A larger amount of helium is found in natural gas – anywhere from about 1 per cent to 7·5 per cent. Helium is totally inert; it does not enter into chemical combination and, therefore, does not have the explosive property that makes hydrogen hazardous. Though helium is twice as heavy as hydrogen, it is so much lighter than air that it provides more than nine tenths the buoyancy of hydrogen. That small loss is more than compensated for by the greater safety helium provides.

With other improvements – lightweight, durable new plastics such as polyethylene and Mylar, more sophisticated new instruments and communication systems – aeronauts were able to reach altitudes above 99 per cent of the earth's atmosphere. In these rarefied regions, the helium expands so rapidly that there is a danger that the thin plastic bag will swell too fast and burst like a toy balloon too vigorously inflated, a danger that must be averted by controlled release of the gas as the balloon ascends. There comes a point, of course, as the air thins, where the balloon and the volume of gas it contains weighs no less than the air it displaces. There the phenomenon of the lighter-than-air balloon reaches its ultimate limitation. Beyond this point rockets are needed to provide the lift necessary to launch men and their instruments farther into space.

Lieutenant Colonel David G. Simons described how it felt when the balloon he was riding in Project Man High I reached this limit, on 20 August 1957. His craft had been rising rapidly and steadily as if 'on an endless elevator'. Then the rate of ascent began to drop off sharply. The balloon was almost full, a swollen ball two hundred feet wide.

> The altimeter faltered, then stopped. I noted the time: 11:40. It had taken me two hours and eighteen minutes to bump against the earth's ceiling. Slowly the needle dropped, then rose again. The balloon was gently bouncing like a basketball being dribbled in slow motion in an upside-down world. It had ascended as high as it would go.'[11]

At 102,000 feet (about 19 miles) Simons had reached the limits of balloon flight. And yet he had taken only the smallest step into earth's aura. If its depth were the height of an 800-storey skyscraper, Simons would have reached the 35th floor. At that point he was already far beyond the realm of the weather (known as the *troposphere*, from the Greek word *tropos*, meaning a turn). He was high above the mists and the rain, the snow showers and the thin curled cirrus clouds. He was far above the region where tornadoes form and rainbows arch their delicate colours across the sky. He was even above the giant thunderheads from whose tops float streams of ice crystals in shining anvils.

Project Man High had penetrated into the stratosphere. Here, strangely enough, temperatures begin to rise again. It is a remarkable fact, discovered by the aeronauts, that the earth's space environment seems to be layered like the skin of an onion. At an altitude of about 100,000 feet, Simons reported, 'I saw something I did not believe at first. Well above the haze layer of the earth's atmosphere were additional faint thin bands of blue, sharply etched against the dark sky. They hovered over the earth like a succession of halos.'[12]

It is not immediately obvious why the earth's atmosphere is layered. Molecules of gas all move very rapidly and tend to even out any differences in temperature and density; so common sense would tell us that there should be no sharp transitions or layers. We would expect the earth's atmosphere and all the influences caused by the planet's presence to recede uniformly and merge imperceptibly with outer space. This common-sense assumption was the basis of the accepted scientific model describing the distribution of the gases around the earth less than a century ago.[13] It was not until men actually penetrated into the atmosphere with their measuring devices that the complex structure of the earth's aura was recognized. As David Simons observed, it is arranged in a succession of haloes. These haloes all lie closest to the earth at the poles and gradually expand to higher altitudes towards the equator. Their heights also vary a little from day to day and from season to season.

The first layer (the troposphere) contains atmosphere in the form we are familiar with here on earth. It rises to about 6 or 7 miles in the mid-latitudes – 4 or 5 miles at the poles and 10 to 11 at the equator. In this layer the gases do thin out with increasing altitude, as the aeronauts found, and temperatures drop – on the average about 3·6°F or 2°C for each 1,000 feet. Figures as low as −80°F (−62°C) have been recorded near the roof of the troposphere. Then suddenly as higher altitudes are attained the temperature stops falling; it levels out and begins to rise

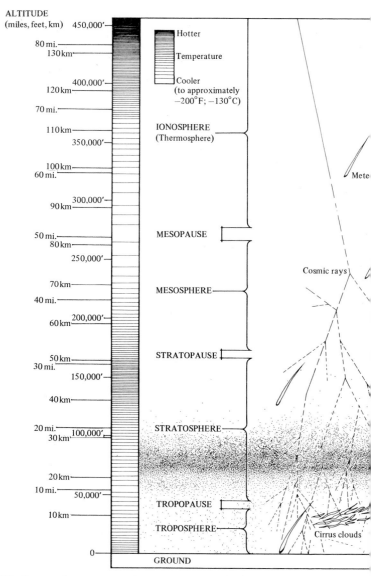

ALTITUDE
(miles, feet, km)

450,000'

80 mi.
130 km

Hotter

Temperature

Cooler
(to approximately
−200°F; −130°C)

400,000'
120 km

70 mi.

110 km
350,000'

IONOSPHERE
(Thermosphere)

100 km
60 mi.

Mete

90 km
300,000'

50 mi.
80 km

MESOPAUSE

250,000'

Cosmic rays

70 km

MESOSPHERE

40 mi.
60 km
200,000'

50 km
30 mi.

STRATOPAUSE

150,000'

40 km

20 mi.
30 km
100,000'

STRATOSPHERE

20 km

10 mi.
50,000'

10 km

TROPOPAUSE

TROPOSPHERE

Cirrus clouds

0

GROUND

Fig. 1. This diagram shows an approximate cross-section of the lowest 85 miles of the earth's atmosphere in the mid latitudes. There are seasonal variations.

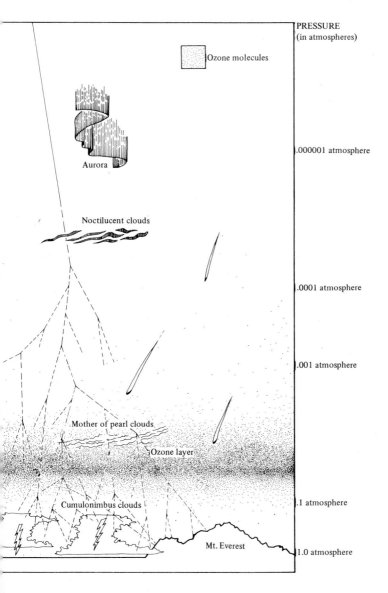

PRESSURE
(in atmospheres)

Ozone molecules

Aurora

Noctilucent clouds

.000001 atmosphere

.0001 atmosphere

.001 atmosphere

Mother of pearl clouds

Ozone layer

Cumulonimbus clouds

.1 atmosphere

Mt. Everest

1.0 atmosphere

slowly. The place where this change occurs is called the *tropopause*. It marks the dividing line between the troposphere and the stratosphere. Here lie the wide flat tops of the towering cumulonimbus. Here, too, is the cave of the jet winds where currents of air blow at speeds sometimes exceeding 200 miles an hour.

Above this transition point the tenuous gases become more and more highly charged with energy. At 30 miles temperatures as high as 86°F (30°C) have been measured. These hot gases are too thin to have much warming effect on a satellite or rocket that penetrates to these altitudes and there is so little oxygen here that a candle will not light. But widely diffused throughout the stratosphere is a special kind of oxygen which captures the ultra-violet radiation of the sun and transmutes it into heat and chemical energy.

At about 31 to 32 miles high another reversal takes place. Temperatures start to decline again and reach their lowest point at the *mesopause*. It is a curious fact that the coldest temperatures in the world occur about 50 miles over the tropics while the hottest temperatures are found above the frozen arctic lands at even greater altitudes where a wide band of very diffuse and active matter flickers and glows by its own fluorescent light. In mysterious ways this highly charged air exerts an influence on all the deeper layers that lie between it and the crust of the planet.

But the first aeronauts who looked out upon the 'earth's ceiling' from a balloon could see only the faintest hint of the powerful forces that create a many-layered halo around the planet. The air at 100,000 feet was clear with a transparency unknown on earth. The horizon was a brilliant iridescent blue, and straight above, the sky was a dark velvety violet several shades deeper than twilight.

CHAPTER TWO

WATER FALLING FROM THE SKY

We shall walk in velvet shoes
Wherever we go
Silence will fall like dews
On white silence below
We shall walk in the snow.

– Elinor Wylie[1]

Every year in late May the eastern provinces of India lie brown and blistered like a giant pie crust baking under the heat of an unclouded sun. Every drop of extra moisture has been sucked from the surface of the soil, drawn up into the relentless blue sky. The wide, meandering river-beds are funnels of gravel and sand. The rice fields are cracked and desiccated, the grasses have turned brown. Even the leaves of the ancient peepul trees have wilted, providing scant shade for the swarms of monkeys who crowd their gnarled branches and the children who play naked in the dusty streets.

Then suddenly one day shadows begin to fall across the land; great cloud masses move in from the south-west. Like stampeding flocks of frightened sheep, they pile up against the high Himalaya range that rises steeply to the north and east. Pressed by the winds that now stream in across the land, bearing moist air off the Indian Ocean, the clouds crowd against the towering mountainsides until finally, finding no other escape, they scramble up the steep slopes towards the shining summits of ice and snow. At these higher altitudes the air expands and cools; it can no longer carry so great a burden of moisture. The clouds spill their rain in a torrential stream that may go for forty or fifty days and nights without ceasing. Within a week or so the land cannot soak up the quantities of rain that fall. Water stands in the fields. The once dry river-beds become violent torrents; the hard-baked clay of the village streets turns into streams of yellow mud. The small mud-and-brick houses become waterlogged inside and out. Clothes mildew, wood swells, and it is now as impossible to find a dry spot as it had been to find a cool moist place just a few weeks ago. Even frogs flee the rain-soaked fields and seek out the driest corners of the houses.

In the province of Assam in the valley of the Brahmaputra there is a little town named Cherrapunji, which has an average rainfall of 424 inches a year – about thirteen times as much as falls on Illinois – and this rainfall comes almost entirely during the summer months. But in India this astonishing phenomenon occurs with clocklike regularity almost every year; so many forms of life – even mankind – have become adapted to it. Animals and people rejoice when the rains come. The water buffaloes wallow in the flooded fields. Men, women, and children work ankle deep in mud, planting the rice seedlings. Each year they pray that the rains will continue until the rice has matured and will provide a good harvest.

A rainfall of several hundred inches in fifty days, which is such a blessing in India, would completely paralyse life in many other parts of the world. During twenty-five days in January 1937, fourteen to nineteen inches of rain fell throughout much of the midwestern United States, causing the most widespread flood in American history. It destroyed life and property in eleven states. A *New York Times* reporter who flew over the flooded area described the scene of devastation along 700 miles of the Ohio Valley in these words:

> Louisville looked like a new Atlantis with from three-fourths to four-fifths of its total area almost eavedeep in dark brown water. Cincinnati, except in a few high spots, seemed planted in a chocolate sea. Near Evansville, Indiana, where the Ohio seldom broadens out more than a half mile or quarter mile, the waters had spread from fifteen to twenty miles beyond the river banks. The river seemed to cover a dead world, with barnroofs and treetops and farmhouses barely visible above the tide.[2]

Because this flood developed very slowly over a period of several weeks the loss of life was moderate but the other costs were enormous. The final statistics on this flood totalled 137 dead, about 1,000,000 homeless, and $417,685,000 in property damage.

The survivors of the great American flood and the inhabitants of Cherrapunji would find it hard to believe that if all the water which is present in the earth's atmosphere were condensed and spread evenly over the surface of the planet it would only make a puddle a little over one inch deep. This fact seems hardly credible when we consider that forty inches can fall on one place in twenty-four hours. The explanation of this paradox, of course, is the uneven distribution of the water contained in clouds, mist, and vapour. There are some places in the world that receive no precipitation at all and others that have negligible amounts.

I remember being told when I visited the peninsula of Baja California that years might go by without any rainfall. The extreme barrenness of the land did testify to the truth of this statement, but during the week I was there, the skies clouded up and one day it began to rain. It was not the sort of precipitation that would even be mentioned in the weather reports in the midwestern United States. One drop fell on every square foot or so of the sandy, barren ground, but the local people were overcome with amazement. They stood in clusters staring up at the clouds. Mothers brought their small children out of the little conical thatched houses and pointed up at the heavens. '*El agua!*' they exclaimed. '*El agua se esta cayendo del cielo!*' Water is falling from the sky!

Water falling from the sky is one of the most important of all the phenomena that make life possible on land, because all life depends upon water. It is the principal ingredient of living tissue. Our bodies are three fifths water and similar proportions are characteristic of most animals and vegetation. Water acts as a transportation system for many of the other chemicals which comprise living tissue. Most of these chemicals are soluble in water and can be conveniently carried by this fluid medium from cell to cell, assuring a continuous stream of nourishment and energy. The fact that water is essential is one of the reasons why scientists believe that life began in the sea and only many aeons later moved onto the land. If the earth's atmosphere did not contain water vapour which condenses into rain or snow and falls on the land, living creatures which have evolved in a water medium could not have survived in this new land-based environment.

Taking together the seas, the lakes, the rivers, the air with its film of clouds, we can think of the planet as being encased in a gigantic bubble – a thin membrane, part liquid, part vapour, constituting the wet warm matrix for living things. If all of this water were poured into a giant container it would occupy 53 billion cubic feet. Most of it is in the seas. Fresh water comprises only 3 per cent of the total and three fourths of that is locked up in polar ice caps and glaciers. Almost all the rest is found in lakes and springs and ground water. Only one thousandth of 1 per cent is present at any one time in the earth's atmosphere.[3] And yet this tiny proportion is responsible for many of the most beautiful, the most frightening, and the most depressing effects that mark the cycle of a typical year on earth. It makes the dew that collects like strings of pearls on meadow grass. It makes the cold driving rain on a November day, the shining white perfection of snowflakes swirling softly on the winter wind. It makes the fogbanks that roll in without warning over the ocean shore, and it provides the energy that generates the hurricanes, spins the tornadoes, and accelerates the blizzards that sweep down from the polar regions over the temperate zones.

Water is a most remarkable substance. It is made of a very simple molecule, containing just three atoms, two of hydrogen and one of oxygen. In contrast, the molecule of ordinary table sugar contains forty-five atoms and a simple protein molecule contains two or three thousand atoms.

Water has several special properties that are very advantageous in its role as a matrix for living things. It is a liquid within the temperature range most suited to life processes. In this liquid form it changes temperature very slowly. More heat is needed to raise the temperature of a given quantity of water than is necessary for any other known liquid. Conversely, as heat is dissipated, the temperature of water drops very slowly. This gradual warming and cooling characteristic produces a remarkably stable and benign environment. The proximity of water in lakes and oceans and in the atmosphere helps to modify the temperature extremes that occur on land.

Like all other liquids, when water is cooled it contracts, but just before it actually solidifies into ice, water does something very unusual – it expands. Because of this strange property, the coldest water is less dense and remains on the surface of lakes and rivers. They freeze on top first and the shield of ice serves to protect the deep layers from contact with the colder air; so most large bodies of water do not freeze solid.

When ice melts a certain amount of heat must be supplied and this is known as the *heat of fusion*. When water freezes into ice this same amount of heat is given off. Water's heat of fusion is large compared to other liquids.

The evaporation of water also requires the addition of heat energy from the environment. This effect is familiar to us all – after we climb out of a swimming pool our wet bodies become cold because the water evaporates, taking the *heat of vaporization* from the air immediately next to our body surfaces. A wind serves to speed evaporation and increase the chilling effect.

Water that has evaporated is a gas, composed of single molecules free to move independently among the other molecules of the atmosphere. This water vapour, like many other gases, is entirely invisible. The separate molecules are so small and widely spaced that light passes by them without being significantly deflected by their presence and the air that contains them may be perfectly clear and transparent. In this gaseous state water is pure; its molecules are not linked with any other molecules. Salt and all the other chemicals that are dissolved in sea water, for instance, are left behind when water is evaporated from the oceans; so the transformation of water into vapour is an efficient purifying process.

When the air is warm, it can hold more water in the vapour form than it can at lower temperatures. As the air cools the water molecules tend to condense back into the liquid form, making tiny droplets of water which are still small and light enough to be airborne, suspended by the rapidly dancing molecules of the gases in the atmosphere. Fifty billion of these droplets would not fill a tea cup; but they are the substance of all the clouds in the sky, from the filmy curtains of cirrus to the towering cumulonimbus.

The mists that veil mountain valleys on summer mornings, the dew and the frost that decorate the low-lying vegetation at dawn, are all manifestations of this condensation of moisture from air that has been chilled. On clear nights the ground loses heat rapidly and the air next to it is cooled to the point where condensation occurs. Mists and fogs are simply clouds at ground level. They collect in little pockets in the terrain because cool air is heavier than warm air and tends to seek out low places.

In some of the most arid lands in the world people have learned to utilize almost every drop of moisture that condenses from the atmosphere. An ingenious ancient custom practised in desert lands is the construction of shallow pits lined with rounded stones surrounding a

3. Early morning mists.

favoured tree or vine. The stones cool off rapidly at night and dew collects on them, trickling down to the roots of the plant. This method provides enough moisture to sustain an olive tree in climates where there is scarcely any rainfall.[4] In some locations near the seashore – for example, Tasmania and the Hawaiian Islands – fog contributes as much as ten or twenty inches of moisture per month to the land.[5]

The condensation of water vapour into dew or cloud droplets is accompanied by the release of the heat of vaporization, which warms the surrounding air. When this process occurs at ground level as in the formation of dew or ground fog, the heat serves to reduce the chilling effect of the cooler earth surface. As the sun rises and warms the air, the fog and dew are soon evaporated again, completing the cycle. But under certain circumstances the condensation of water vapour can cause a much more dramatic effect. If air is warmed by the sun and contains a large amount of moisture in the form of water vapour, it rises. This occurs for two reasons – because it is warm and because moist air is actually lighter than dry air. (A cubic foot of water vapour weighs about two fifths less than a cubic foot of dry air at the same temperature and pressure.) As the air reaches higher altitudes it expands and cools; and the water vapour begins to condense into cloud droplets. This process releases heat and the air rises again. Borne aloft to even colder regions of the troposphere, the water droplets are chilled to the point where freezing takes place, forming crystals of ice and snow. The freezing again releases heat energy and the column of air rises still higher. This self-augmenting process enables currents of moist air to defy gravity and rise twenty or thirty thousand feet above the earth's surface, carrying water first as a gas, then as droplets, then as snow or ice crystals that form at the top of a thunderhead. A storm has been created, a storm charged with fantastic amounts of energy squeezed from countless billions of those tiny simple molecules of water.

For many years meteorologists were baffled in their attempts to answer a fundamental and seemingly simple question: How do raindrops form? Raindrops are at least a million times bigger than cloud droplets. Both theory and laboratory experiments showed that cloud droplets would not tend naturally to come together into larger drops.

The process of freezing proved to be the clue that solved part of the puzzle. As temperatures drop below 32 °F (0 °C) most of the cloud droplets remain in the liquid state. In fact, temperatures can fall very far below 'freezing' without any crystallization taking place and the moisture-laden air is said to be *supercooled*. Occasionally, however, one in a million of the droplets does freeze and once the tiny ice crystal exists in the supercooled cloud, water vapour molecules are attracted to it. Building outwards from the centre, the molecules are added one by one. As though in a miniature game of Farmer in the Dell, each molecule attracts others and the pattern is built up symmetrically in ever widening circles. All this occurs so rapidly that to our senses it appears to be instantaneous. Each crystal, an original masterpiece of lacelike perfection, suddenly materializes from what seems to be empty space.

Although it is known that the presence of ice starts this process, there is still no scientific theory that adequately explains why thousands of individual molecules all at once forsake their separateness and join together with mathematical precision to form an organization of the highest complexity and order. The birth of a snowflake is a dramatic demonstration of the unknown forces underlying all nature. As Loren Eiseley describes it:

> Water has merely leapt out of vapor and thin nothingness in the night sky to array itself in form. There is no logical reason for the existence of a snowflake any more than there is for evolution. It is an apparition from that mysterious shadow world beyond nature, that final world which contains – if anything contains – the explanation of men and catfish and green leaves.[6]

As the single ice crystal grows it becomes heavy enough to fall through the cloud, colliding with cloud droplets. Tiny splinters break off the crystal, leaving a trail of ice nuclei to start the formation of new snowflakes. These grow and fall and splinter, spawning more flakes, and so a flurry blossoms from the formation of a single crystal.

This explanation pushes the mystery back one stage, but it is still necessary to explain how and why that first crystal forms. It turns out that microscopic bits of matter in the cloud – particles of dust or sand or salt – act as nuclei around which freezing takes place. In supercooled clouds the introduction of a few particulates triggers a chain reaction and precipitates the snowstorm. If no particles of matter are present at all, water vapour can be cooled to $-40°F$ ($-40°C$) before freezing occurs spontaneously.

Woven of air and water in its purest state, the snowflake is incredibly light. A typical powder snow has twenty parts of air to just one of water, but some very gossamer snows have been measured in Finland and found to contain as little as one part of water to three hundred parts of air.[7]

Being so delicate and airy, many snowflakes never reach the ground at all. They float high in the upper reaches of the troposphere, swirled by wind currents, forming the luminous white mare's tails and the iridescent curls of the cirrus clouds. Those flakes that become heavy enough to defy the winds start to fall towards the earth, gathering other flakes to them as they descend. In the winter portions of the globe where all the air is below freezing, the flakes flutter slowly downwards, finally reaching the earth's surface as snow. Most of the flakes that reach the ground are clumps of fragmented ice crystals. [See Plate 4.] The single six-sided snow crystal is relatively rare. It can assume an almost infinite variety of exquisitely detailed forms.

4. A snowflake, its fragile beauty still intact, settles towards the black ice of a pond in Vermont.

One of the most notable characteristics of a snowstorm is its stillness. The white flakes drift and float noiselessly through the air. The downy white blanket that covers the ground mutes all sounds. Even the strident noises of the city are softened. This quietness is not just a subjective impression; it is based upon a physical fact. The snowflakes act like little acoustical baffles, breaking up the sound waves and absorbing them just as though they were tiny floating bits of down. And as a down comforter would, the snow cover holds in the earth's heat. The temperature of the ground under the snow is warmer than the land that is exposed to the winter winds.

On the other hand, snow cover has quite an opposite effect on the air. It is a very efficient reflector of radiation, so that 80 to 90 per cent of the energy of sunlight that falls upon it is reflected. After a good snowfall we are apt to experience a period of severely cold weather because the sun's energy is reflected instead of being absorbed by the earth.

In the atmosphere above the portions of the earth that are turned towards summer, the water molecules undergo a more complex series of adventures. Some of the snowflakes that are formed at high altitudes start to fall, encounter warmer air, and turn into raindrops. Rain falls more efficiently than snow because it is denser. On their downward path the raindrops grow by collecting many of the little cloud droplets that they encounter. The largest amount of precipitation on earth occurs in this form.

Occasionally, too, raindrops grow directly from cloud droplets when some unusually large particulates are present. The droplets forming on these nuclei are large enough to attract neighbouring cloud droplets to their surface and coalescence occurs.

In the very turbulent clouds such as the great thunderheads, raindrops may be caught up in the strong updrafts that are characteristic of these storms. Swept back up into the very cold regions of the atmosphere, the raindrops freeze, not as finely constructed snowflakes, but as tiny pellets of ice. Where the updraft is strong enough these little pellets dance on the air like Ping-Pong balls on a jet of water. At great altitudes – perhaps 25,000 feet above sea level – the strength of the updraft weakens; the ice pellets are dispersed in all directions, falling down on the outside of the central updraft. Many such ice particles falling together cause a cool downdraft. These strong opposing currents so close to each other in the cloud add to the turbulence. The ice pellets are carried in a wild zigzag dance through the storm. As they pass through supercooled clouds, droplets adhere to the ice pellet, freezing into another layer of ice. Sometimes the ice pellets are carried repeatedly upwards and down again in the opposing currents of air, growing to great size and emerging as hailstones built up of many layers of ice. Hailstones are not the symmetrical balls one might expect. They are so different in character from the perfection of the snowflake that it is hard to believe they are both frozen forms of the same element. Hailstones are distorted, odd-shaped objects, reflecting the violent, storm-tossed conditions of their formative growth.

A strange lighting phenomenon usually accompanies a hailstorm – an eerie greenish glow bathes the sky and countryside. This striking effect is associated in popular thought with tornadoes, and since hailstorms and tornadoes do frequently come together, there is some justification for this belief. But the green light has often been seen during storms when no tornadoes are present and pieces of ice shower down out of clouds illuminated with this phosphorescent glow, whose source has not been explained.

Hailstones come in many sizes. Most are smaller than peas. Some are as large as grapes and these can cause widespread destruction, turning ripe fields of corn into acres of rubble and stripping fruit, leaves, and even the bark from trees. Occasionally very large hailstones the size of oranges or grapefruit are pelted at enormous speeds (up to a hundred miles an hour) from a thunderstorm, killing people and animals. Hundreds of people died in a particularly disastrous hailstorm in India in 1888. In 1360 an English army led by King Edward III lost so many men and horses in a hailstorm outside Paris that the king signed the Peace of Bretigny.[9]

The destructive power of hail has long been feared and men have dreamed of finding some way to control this violent phenomenon. In centuries past when a storm threatened vineyards loaded with ripening grapes, church bells were rung to ward off the hail. It was believed that the vibration of the sound waves could influence the formation of hailstones. This principle was carried a step further in the seventeenth century when cannon were fired into storm clouds to protect the grape harvest in the Beaujolais wine district of France.[10]

Around the turn of the twentieth century an enterprising Italian invented a gun which fired a blank charge, blowing a doughnut of hot air up into the sky. These guns, which looked like enormous megaphones supported on three wheels, were exhibited in Padua in 1900 and were so successfully promoted that 2,500 of them were in use before it was finally demonstrated that they had no measurable effect on the size and force of hailstorms.

Efforts to reduce the damage from hail still continue. Recently scientists in the Soviet Union reported that they had developed a hail-suppression project that reduced hail damage by 80 to 90 per cent.[11] Other experiments are being conducted in France, Germany, Italy, Argentina, and the United States. The techniques all make use of a principle discovered in 1946 by two American scientists, Vincent Schaefer and Irving Langmuir. During the Second World War they were investigating the cause of ice formation on airplane wings, and in the course of this research they found that many clouds at high altitudes are supercooled. Anything passing through these clouds (an airplane, for instance) can act as a trigger to start freezing the water in the cloud droplets. Schaefer followed up this discovery with laboratory experiments. He found that by breathing into a freezer he could make miniature supercooled clouds. Then by introducing a very cold metal rod or particles of dry ice, he could turn the cloud into snow and ice crystals. On 13 November 1946, Schaefer conducted a field experiment to test his theory. Flying in a small plane, he identified a supercooled cloud at 14,000 feet over Schenectady, New York, and scattered six pounds of dry ice into its centre. The cloud, approximately four miles long, turned into a snowstorm.

Schaefer's method was based on the principle of chilling a few cloud droplets to extremely low temperatures by introducing dry ice, which is frozen carbon dioxide and has a temperature of $-110°F$ ($-79°C$). One of Schaefer's colleagues, Bernard Vonnegut, suggested that the same triggering action might be achieved more economically by using crystals that resemble ice. Silver iodide crystals were tried and, as Vonnegut had predicted, the droplets in the supercooled clouds froze around these crystals. The silver iodide does not chill the water any further, but because it is similar in shape to the ice crystal, it acts very effectively as a nucleus for crystallization. As each crystal grows and shatters, a chain reaction is set up that rapidly turns the whole cloud into snow.[12]

The important thing about a trigger reaction of this kind is that extremely small quantities of the added material can result in impressively large effects. Irving Langmuir estimated that a single dry ice pellet the size of a green pea falling through a cloud two thirds of a mile in thickness could produce a hundred thousand tons of snow. The heat released by this amount of water being turned into snow would roughly equal the heat liberated by an atomic bomb such as the one that destroyed Hiroshima.[13]

The trigger effect makes it possible for man to dream of controlling the weather. By scattering a few pounds of crystals it may be possible actually to channel and divert the enormous forces that build up in the deceptively delicate-looking clouds that circulate around our planet. Rains can be encouraged to fall on areas suffering from drought. Clouds that are generating hail may be caused to precipitate more rapidly so that many little hailstones take the place of a few large destructive ones. These possibilities are all being vigorously pursued. The basic principles seem simple at first glance, but when applied to an actual weather situation they are extremely complex. Meteorologists are becoming increasingly aware of the inadequacy of the theoretical basis for large-scale weather modification.

In 1972 scientists at the National Center for Atmospheric Research set up a programme of hail-suppression experiments in a region known as 'hail alley', a 625-square-mile area on the Colorado–Nebraska border. They monitored about thirty storms a year, seeding half of them and leaving the other half as the control group. After several years, when the results were analysed statistically, no significant reduction of hail was found. Indeed, there was some evidence that in certain types of storms seeding may have increased the destructiveness of the storm. These negative results have cast doubt on the previously accepted theories of hail formation and suppression and have inspired the participating scientists with a fresh awe for the complexities of atmospheric phenomena.[14]

5. *The result of cloud seeding. Air Force Cambridge Research Laboratories scientists made this more than three-mile-wide hole in a supercooled cloud deck in one hour.*

In the meantime weather-modification projects continue in many places around the world, as men exploit the newly discovered triggering action in an attempt to reshape the weather patterns to suit their particular needs. In South Dakota almost a million dollars was spent each year from 1972 through 1975 for hail-suppression and rainmaking flights to augment the marginal precipitation over the wheat fields of the High Plains. The cost, borne by the state and counties, averaged about three and a half cents an acre. During 1976 a similar programme was carried out by twelve of the counties without any financing from the state. In the Soviet Union and Yugoslavia thunderclouds are scanned by radar, and when one is found that appears to be capable of producing large hailstones, rockets are aimed into it at the level where supercooled droplets accumulate. Each rocket seeds the cloud with a spray of iodide crystals. Great success has been claimed for these projects, but many meteorologists are sceptical. It is hard to differentiate between a streak of luck and a really positive result from these projects.[15]

The fact that little causes can precipitate large-scale changes in the atmosphere can be a disadvantage as well as an advantage to man. Over the past few decades human activities have brought about inadvertent changes and added many triggering substances to the earth's atmosphere. Thousands of tons of dust and ash are poured forth every day from industrial smoke stacks and exhaust pipes of automobiles. Many of these tiny particles can act as nuclei for condensation just as the silver iodide crystal does. The smog and haze that have become characteristic of cities are caused by condensation of water on these small particulates in the atmosphere. In air polluted by industrial waste there are often as many as a million particles per cubic inch. Weather records provide strong evidence that the presence of this large number of particles is altering the weather. Measurements of high cloud cover over the United States have shown a significant increase since 1965.[16] At La Porte, Indiana, the average annual precipitation is 50 per cent greater since the installation of steel mills, foundries, and refineries at Gary, upwind of La Porte.[17]

The La Porte situation was first documented in the early 1950s. About twenty years later a much more ambitious study of the way cities modify the weather was undertaken in the St Louis area. This city was chosen as the subject of the research project because it is believed to be large enough to cause significant effects and is located in the centre of relatively flat land far from any large body of water. Measurements can be conveniently made in a wide circumference around the city. For about five years extensive data have been collected on temperature, rainfall, hail, sunlight, and the levels of various pollutants. The results are very interesting. The rainfall in an area ten to thirty miles east of St Louis (in the direction of the prevailing winds) is 20 per cent higher than the rain in the city itself or in the outlying regions to the west, south, and north. Furthermore, the additional precipitation comes mostly in the form of heavy downpours of one inch or more. The incidence of hail is also much higher.

The scientists conducting these studies are cautious in their statements because they are well aware of the many variables that affect the weather. However, there is a general consensus that these anomalies are probably caused by the presence of the city. The temperature in the city averages three to five degrees above that of the surrounding countryside. Concrete and masonry absorb heat efficiently during the daytime and release it very slowly at night – much more slowly than a similar area of bare ground or ground covered by vegetation. This island of heat causes a rising column of air over St Louis and, since it is heavily laden with man-made particulates, there are plenty of nuclei to trigger the formation of rain in the resulting clouds.

Another factor which contributes to the total effect of the city is the obstruction that it offers to wind flow. The tall shapes of the buildings project into the air stream, slowing the natural currents. The scientists found that wind speeds across St Louis were reduced as much as 40 per cent, thus leading to more stagnant air conditions and allowing the updraft from the heat island to predominate. For these reasons city-made clouds tend to rise to greater heights than those generated over the surrounding countryside. They seem to breed more violent rainstorms and increase the incidence of hail.[18]

From the results of this study in one typical American city, it is clear that inadvertent climate modification is probably very widespread. Much more information needs to be collected from similar sources in different climatic zones in order to understand in more depth the effect of man's activities on cloud formation and on the distribution of water around the world. For example, the rate of evaporation of water from the land is being reduced every year by the paving of large areas of the earth's surface. On the other hand, an increasing amount of water is being distributed by irrigation projects to dry desert areas where evaporation proceeds very rapidly. Do these changes balance each other? We really do not know. The slightest alteration in the relative rates at which water is drawn up from the land and the sea and distributed again as rain and snow could cause very important changes in our environment. When we consider how small is the proportion of the earth's water supply that is present at any one time in the atmosphere and how vital it is to the very existence and quality of life on earth, we can appreciate the need for monitoring this balance with the greatest care.

An instinctive recognition of the fundamental importance of water flows deep within all of us. Many people hold water sacred. They are baptized in it; they pour it over their heads at sunrise; they touch it to their foreheads in prayer. The pleasure of watching waves break on a sandy shore or listening to the rain on the roof is not just sensual. It is as reassuring as the presence of a mother is to her child. Water is there, providing our nourishment, cleansing our bodies, wrapping us in a cool, misty embrace, and singing a lullaby of a time long ago when all life was cradled in the sea.

CHAPTER THREE

SPINDRIFT OF THE OCEAN OF AIR

A sky without clouds is a meadow without flowers, a sea without sails.

- Henry David Thoreau[1]

At dawn today clouds are rising from the surface of Lake Michigan, tiny wraith-like plumes that spiral upwards from the waves, the fingers of their upstretched arms catching the first rays of the rising sun. The air is extremely cold this January day; the sky overhead is clear and a light wind is blowing from the west. In the east the horizon is obscured by a bank of clouds which the sun has outlined in fiery red and gold. These clouds are heaped in a continuous low bank with ragged tops extending into the blue; their massed shape echoes the form of the drift of tiny clouds that stream upwards from the water. As the sun climbs above the cloud bank it turns the smoking surface of the lake into a luminous air-seascape that moves and flows with dawn colours.

'Sea smoke' this effect is called in the arctic, where it forms on the edge of the ice pack. But it also occurs in many other places, wherever large bodies of open water are exposed to very cold air. The temperature differences between water and atmosphere may be as much as forty degrees; so the atmosphere in contact with the water is warmed and takes on an extra load of water vapour. The warmth causes it to rise into the much colder general air mass, where it is chilled again and gives up its extra moisture in a little plume of condensation. Miles out from the shore of the lake the warming effect is more pronounced. The clouds are larger; they rise higher and form the bank along the eastern horizon.

The smoking of this witches' cauldron takes place most frequently at night when the air drops to its lowest temperature and the winds are light, but the most dramatic visual effects usually occur at dawn. During the day the sun warms the air. The plumes do not rise as high but flocks of tiny puffy white clouds can be seen nestling close to the body of the lake. Looking down at the water surface from a tall building or a high bluff, you will notice that these cotton puffs scattered on the blue are strikingly similar to clouds seen in views of the earth from space. Only the scale is different.

All clouds, from the tiniest wisps of fog to the solid layers that blanket thousands of miles, mark the places where water vapour can no longer hide in the sky and by their very presence – by their constantly shifting distribution, their infinite mobility of form – they draw a continually changing diagram of the atmosphere's dynamic flow. Astronauts looking back on our planet from outer space have all been struck by the unexpectedly large amount of cloud cover, moving in a restless swirl across oceans and continents.

From earth-side, looking up, men have tried for thousands of years to understand the messages written in the clouds. Are they harbingers of destructive storms or will they bring the anxiously awaited rain? It was noticed very early that certain types of cloud formations were usually associated with particular weather changes. The first step in understanding these relationships was the classification of clouds.

The cloud names commonly used today are derived from Latin: *cumulus* means a heap or pile; *stratus*, a layer; *cirrus*, a curl; *nimbus*, rain; and *alto*, raised or elevated. The first three designate the principal cloud types and the last two are used as qualifying terms. Within this small vocabulary the myriad cloud effects are characterized. *Cumulonimbus* means the great heap of cloud that produces a rainstorm; *altostratus* is a high uniform layer of cloud; and so on. There are many possible combinations, and fine distinctions exist between the various types.

Usually condensation occurs whenever air is cooled to the 'dew point' – the temperature at which it can no longer contain as water vapour the amount of moisture that it is carrying. This point varies with the amount of water vapour present in the air. Very dry air over the desert must be chilled to about 35° or 36°F (2°C) before condensation takes place. Since this temperature drop rarely occurs, the desert skies are almost always clear. On the other hand, extremely humid air may begin to turn into a cloud when the temperature is as high as 80°F (26°C). The air temperature may be so close to the dew point that the balance can be tipped by just a degree or two. In this situation clouds form and dissipate right before our very eyes like phantoms materializing out of thin air and then, without a word of warning, becoming invisible again.

6. Cumulus clouds.

Cumulus clouds are the ones that form most commonly in this delicately balanced atmosphere. Plumes of hot air rise from warm land or water into cooler air, like the wisps of steam rising from Lake Michigan on a cold day. Over the tropics this phenomenon is multiplied in size manyfold. The plumes of rising air carry large amounts of moisture. Fluffy white cumulus clouds appear by midday in hot humid climates and build up throughout the afternoon, only to dissipate as the sun goes down and no longer heats the face of the planet. Satellite pictures usually show a belt of these clouds around the waist of the world and they typically occur in clusters. Certain regions such as the vicinity of the Marshall Islands in the Pacific seem to have the optimum conditions for producing them. Meteorologists believe that patterns of air and ocean currents contribute to the clustering.[2]

These great puffs of water droplets are among the most attractive of all clouds – like egg-white meringue beaten from the clear viscous medium of the earth's atmosphere into white froth and scattered on the winds. Their sharply delineated flat bottom edges occur at the condensation level, usually about two or three thousand feet high. Being relatively thick, these cumulus clouds interrupt and reflect the light. When large numbers of them are massed, they shadow each other and create a fantastic landscape with deep purple-shaded valleys, long avenues of luminous pillars and radiant mountain peaks. They are saturated with colour, shading from delicate pink to amber and gold. But the most dominant hue is a pale creamy colour like pearls mellowed by time.

Human beings never had an opportunity to see these wonderful cloud landscapes until they ventured into the skies in balloons and airplanes. From the ground the view of the clouds is limited; many of the most spectacular light effects are blocked by the clouds themselves. But borne aloft to the upper edges of the troposphere, we can now explore the tops of the clouds and can see the whole dazzling panorama brilliantly illuminated by unobscured sunlight. The first men who looked out on this dazzling new world were overcome with awe and delight.

In September 1817, Prince Pückler-Muskau was taken for a ride in a balloon owned by the Frenchman M. Reichhard. The German prince described the experience, which he considered to be one of the high points of his life:

> No imagination can paint anything more beautiful than the magnificent scene now disclosed to our enraptured senses. The multitude of human beings, the houses, the squares and streets, the highset towers gradually diminishing, while the deafening tumult became a gentle murmur, and finally melted into a deathlike silence.
>
> The earth which we had recently left lay extended in miniature relief beneath us, the majestic linden trees appeared like green furrows, the river Spree like a silver thread, and the gigantic poplars of the Potsdam Allee, which is several leagues in length, threw their shadow over the immense plain. We had probably ascended by this time some thousand feet, and lay softly floating in the air, when a new and more superb spectacle burst upon our delighted view.
>
> As far as the eye could compass the horizon, masses of threatening clouds were chasing each other to the immeasurable heights above,

and unlike the level appearance which they wear when seen from the earth their entire altitude was visible in profile, expanding into the most monstrous dimensions – chains of snow-white mountains wrought into fantastic forms, seemed as if they were tumbling headlong upon us. One colossal mass pressed upon another, encompassing us on every side, until we began to ascend more rapidly and soared high above them, where they now lay beneath us, rolling over each other like the billows of the sea when agitated by the violence of the storm, obscuring the earth entirely from our view. At intervals the fathomless abyss was occasionally illuminated by the beams of the sun, and resembled for a moment the burning crater of a volcano; then new volumes rushed forward and closed up the chasm; all was strife and tumult. Here we beheld them piled on each other white as the drifted snow; there in fearful heaps of a dark, watery black, at one instant rearing towers upon towers, in the next creating a gulf, dashing eternally onward, in wild confusion.

I never before witnessed anything comparable to this scene even from the summit of the highest mountains; besides, from them the continuing chain is generally a great obstruction to the view, which, after all, is only partial; but there there was nothing to prevent the eye from ranging over the boundless expanse.

The feeling of absolute solitude is rarely experienced upon earth, but in these regions, separated from all human associations, the soul might almost fancy it had passed the confines of the grave. Nature was noiseless, even the wind was silent; therefore, receiving no opposition, we gently floated along, and the lonely stillness was interrupted only by the progress of the car and its colossal ball which, self-propelled, seemed like the rockbird fluttering in the blue ether.[3]

Although cumulus clouds look invitingly soft on the outside, experienced pilots know that the largest ones are chaotic tunnels of turbulence inside with ascending and descending air currents immediately adjacent to each other. Between the two currents lie sharp bands of high wind shear. Aeroplanes can sometimes pass unharmed through these clouds while others only a few hundred feet away are torn to pieces. During the Second World War a squadron of eight bombers heading north from Australia flew into a line of cumulonimbus clouds which extended for a hundred miles. Only two flew out.

Under every cumulus cloud there is an updraft that feeds warm moist air into the cloud. Glider pilots very early discovered how to take advantage of these natural elevators to gain the altitude needed to continue their flight, but after a number of fatal accidents they learned to use these thermals with caution. The innocent-looking cumulus may be growing into a cumulonimbus with very powerful thermals which can sweep the defenceless sailplane straight up to the top of the developing storm.

Such a tragic event occurred during one of the early gliding contests in Germany. Leo Loebsack, a German science writer, recounted the story:

Between Werra and Fulda, in Central Germany, lies a range of mountains, the Rhön, which recalls wonderful gliding contests, effortless flight across fields and meadows and the marvellous 'thermals', currents of rapidly rising air which can carry the glider to endless heights. During favourable weather the Rhön is a veritable glider's paradise, but on a close summer day in 1938 high above the hillside a tragic misadventure occurred.

Another contest had been announced. Enthusiastic competitors arrived with their streamlined 'birds' of all colours and types. The slightly thundery atmosphere gave promise of record-breaking attempts, and in fact heights of over 26,000 feet were reached in several cases. Then five daring contestants flew into a thunder cloud.

It is possible that they did not correctly read the danger in the cloud's appearance, or that they were carried away by the spirit of competition and got too close. All five of them were suddenly sucked into the center of a violent squall. It jerked the gliders upward and the dense cloud blotted out all visibility. Fearing they would be thrown against a cliff, or perhaps because their gliders were already damaged, the flyers jumped and pulled the release cord of their parachutes.

The consequences were dreadful. Instead of falling gently downwards the parachutes were filled to bursting point by the wind and carried upwards. Higher and higher they soared into increasingly colder layers of the cloud. However hard they tried to steer with their arms and legs they could not escape the howling force of the gale. Huge raindrops soaked their bodies in a few seconds. Hailstones lashed their faces. It was as if nature were punishing this audacious handful of men for venturing into forbidden heights.

We do not know all that happened between earth and sky during these frightful minutes, as only one man, severely injured, escaped with his life. We can only imagine the ordeal of the other four. At a height of thirty-five, forty-five or even fifty thousand feet they must have been enclosed in a casing of frozen water, tossed about like living icicles, stabbed at by lightning, until the cloud released their four lifeless bodies.[4]

The treacherous currents of air that feed the thunderheads exist above the clouds as well as below them. In Project Man High, David G. Simons, floating in the stratosphere at 70,000 feet, had just closed his eyes for a nap when suddenly he was awakened with a start. The gondola was plunging downward and spinning. 'Great God,' he thought. 'The bottom has fallen out! . . . The balloon must have split!'[5] Tugging a Kleenex from the box at his side, he scrubbed the porthole clean so he could look out. Below and to the south-east he saw a line of very tall thunderheads, reaching higher than he had ever imagined such clouds could rise. Now he realized why the balloon was dropping so wildly. It had been caught in a downdraft generated by a thunderhead below. Then as suddenly as the precipitous fall had begun it stopped and Simons was not – as he had feared – sucked like a toy into the centre of the storm.

The towering cumulonimbus clouds are seething cauldrons of energy which is manifested in violent air motion and in electrical effects that are peculiar to this particular kind of cloud. Lightning flickers intermittently like a defective fluorescent bulb, illuminating the cloud from within, and every few minutes dazzling streaks of lightning sear the air between the cloud and the ground.

Measurements made from balloons and by high-speed photography have revealed some of the secrets of lightning formation, but many aspects of this most common and spectacular atmospheric phenomenon are still not understood. It is known that the top of a typical cumulonimbus cloud has a strong positive charge, the middle and lower portion are mostly negatively charged, and the ground below the cloud takes on a potential which is mainly positive. Like an enormous electrical sandwich, the positive top and bottom layers are separated by the negative filling. The potential between the top and bottom of a thundercloud may be as much as 100 million volts. Meteorologists are not certain just how this tremendous potential difference is built up. In some way the swift air streams serve to separate positive and negative charges which in air molecules are normally very closely balanced and, therefore, neutral. One of the most generally accepted theories is that the ice crystals which form at the top of the cloud are shattered in the turbulence of the air streams and the main central cores of these crystals usually retain more than their fair share of electrons, giving them a negative charge. These heavy negatively charged crystals fall towards the base of the cloud, melting into rain and carrying with them the extra electrons, while the smaller fragments are borne aloft on the winds, concentrating their positive charges high in the storm cloud. Dry air is so resistant to the passage of an electric current that the charges thus separated cannot come together and be neutralized until a very high potential difference is reached. But water vapour in the air improves its conductivity and, since the conditions in the storm are extremely variable, certain portions of the atmosphere become conductive first, allowing the passage of immense sparks across the cloud or between the cloud and the earth.

About one out of ten flashes of lightning reach the ground. The others which dart back and forth within the cloud itself may be just as spectacular but our view of them is partially obscured by the storm. Contrary to popular belief, the main surge of electricity in a lightning flash travels from the ground up to the cloud. First a relatively weak leader stroke zigzags through the air from cloud to ground, seeking out the most conductive path. It is immediately followed by a much more massive stroke returning to the cloud. This action is repeated several times, as has been shown by high-speed photography. The whole series occurs in less than a tenth of a second, so it appears to us as a single flash.

The current of electricity ripping through the atmosphere heats up a narrow channel of air to incandescent temperatures – about 18,000°F (10,000°C). It is the brilliant light from this glowing air that we see as the lightning flash. The intense heat causes a sudden expansion of the air, which, as the heat is dissipated, contracts sharply and produces the clap of thunder. A stroke directly overhead makes an earsplitting crash, but when the stroke occurs at a greater distance the thunder is a long muffled rumble as the sound is reflected off intervening surfaces.

Lightning can take on many forms – the forked, the branched, and the sheet lightning are really all the same phenomenon seen from slightly different perspectives. Occasionally strange shapes occur. 'A string of beads' looks like an ordinary flash which has been divided up into a series of luminous spheres, usually spaced about twenty-five to forty feet apart. Occurrences of balls of lightning have also been reported by many observers. These eerie apparitions, which are often blood red in colour, may be single beads which have become separated from a string. They have been said to pass through keyholes, to travel in circles around a room and disappear up a chimney, giving off a hissing noise and a rancid odour, then finally exploding with a loud bang. Although some authorities dismiss the lightning ball as an optical illusion caused by the lightning flash, others believe that it is a small mass of highly charged gas produced by the electric discharge.[6]

Conditions leading up to a lightning storm can often be anticipated by observing the blue lights which flicker and dance from the pointed ends of flagpoles, church steeples, and airplane propellers. The highly charged surface beneath a thundercloud is discharging itself into the air from sharp-pointed surfaces where the potential is unusually strong. At these places electricity can pass most easily into the air. Called St Elmo's fire, this electrical glow was named after an Italian bishop who lived around the year 300. He was noted for his kindness and became the patron saint of Mediterranean sailors. Ironically, the presence of the flickering blue discharge that bears his name indicates conditions that may increase the hazard of a lightning storm. The very high local potentials on tall pointed objects help to initiate a lightning strike, and if the object is not well grounded – like a lightning rod – the strike may cause fire and fatal accidents. Wooden sailing ships at sea are especially vulnerable to lightning damage. An English meteorologist estimated that between 1810 and 1815 no less than fifty-eight ships were struck, and many of them were severely damaged by this fire from the sky.[7]

Some primitive peoples like the aborigines of Tasmania, who had not learned how to make fire themselves, gathered a burning faggot from forests set ablaze by lightning. They fed and guarded this flame as their most precious possession. If it was allowed to die or was drowned out by rain, the tribe waited and prayed for a lightning storm to rekindle their flame.[8]

It was not until Benjamin Franklin performed his famous kite experiment in June 1752 that the electrical nature of the fire in the sky was demonstrated. Describing the experiment in a letter to Peter Collinson, a Fellow of the Royal Society of London, Franklin said:

Sir,

As frequent mention is made in public papers from Europe of the success of the Philadelphia experiment for drawing the electric fire from clouds by means of pointed rods of iron erected on high buildings, and, it may be agreeable to the curious to be informed, that the same experiment has succeeded in Philadelphia, though made in a different and more easy manner, which is as follows:

Make a small cross of two light strips of cedar, the arms so long as to reach to the four corners of a large thin silk handkerchief when extended; tie the corners of the handkerchief to the extremities of the cross, so you have the body of a kite; which being properly accommodated with a tail, loop, and string, will rise in the air, like those made of paper; but this being of silk, is fitter to bear the wet and wind of a thunder-gust without tearing. To the top of the upright stick of the cross is to be fixed a very sharp-pointed wire, rising a foot or more above the wood. To the end of the twine, next the hand, is to be tied a silk ribbon, and where the silk and twine join, a key may be fastened. This kite is to be raised when a thunder-gust appears to be coming on, and the person who holds the string must stand within a door or window or under some cover, so that the silk ribbon may not be wet; and care must be taken that the twine does not touch the frame of the door or window. As soon as any of the thunder-clouds come over the kite, the pointed wire will draw the electric fire from them, and the kite, with all the twine, will be electrified, and the loose filaments of the twine will stand out every way, and be attracted by an approaching finger. And when the rain has wet the kite and twine, so that it can conduct the electric fire freely, you will find it stream out plentifully from the key on the approach of your knuckle. At this key the phial may be charged; and from electric fire thus obtained, spirits may be kindled, and all the other electric experiments be performed, which are usually done by the help of a rubbed glass globe or tube, and thereby the sameness of the electric matter with that of lightning completely demonstrated.[9]

It is a wonder that Franklin lived to tell this story. When he brought his finger near the twine and the key, he was in great danger of electrocuting himself. Driven by enormous potential forces, lightning can break through small insulating barriers and once the leader dart has found a conducting path, the main strike is inevitable.

About four hundred people are killed every year in the United States by lightning. Figures for the whole world are not available, but it has been estimated that, on the average, 1,800 thunderstorms are taking place over the earth at any one time and hundreds of lightning strokes light the sky.[10] Seen from above (as David Simons reported), they look 'like flashing neon displays, the clouds . . . shot through with sporadic pulses of light that showed up in a gorgeous pattern of puff and shadow'.[11]

7. Stratus clouds.

Stratus clouds are much less dangerous and less attractive visually than the cumulus. Their flat structure is formed when the air moves in a very uniform manner to a location where cooler temperature causes condensation. Stratus clouds are usually found where a warm air mass passes over colder air. The warm air, being lighter, rides up over the cold

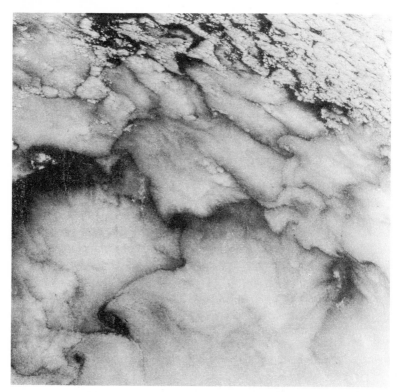

8. Shingles of cellular clouds, each so big that no photograph from earth or aircraft ever showed more than one, roof the western Pacific.

air and its moisture rapidly condenses out in a broad stratus layer of sloping altitude. These clouds generally create a dark grey pall across the sky and, since they bring hours of rain, they are known as nimbostratus.

Sometimes the clouds comprising the layer are broken up into a mottled structure caused by a wave type of circulation within the shallow cloud layer; the air rises up in one place and descends in another and so on at regular intervals. This circulation pattern breaks the cloud layer up into regular strips and produces the typical 'mackerel sky' which is an omen of bad weather. Waves crisscrossing each other at right angles can create the cellular clouds that look like enormous ice floes suspended in an arctic sea. [See Plate 8.]

9. Cirrus clouds.

Stratus clouds occur at many different altitudes from 30,000 feet down to almost ground level. Cirrus clouds, on the other hand, always ride very high in the sky at six to nine miles above the earth's surface. They reflect the warm colours of the sun immediately before dawn and after sunset leaves the rest of the world in darkness. These ephemeral cloud formations are composed entirely of ice crystals and are so tenuous a form of matter that they contain only a few ice crystals in each cubic inch. They are too insubstantial to block out the light or to cast any shadows. The beautiful shining filaments and mare's tails that describe curved patterns across the sky are ice cascades seen from underneath as the crystals fall through the thin air of the upper troposphere. Because

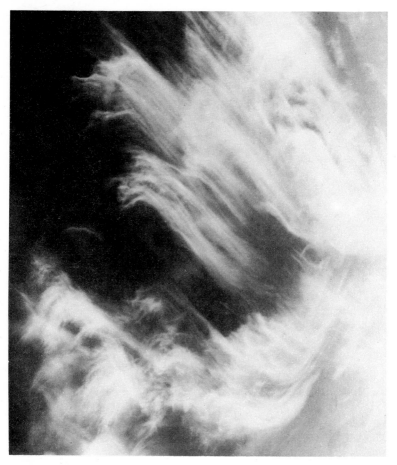

10. Cirrus clouds composed of cascades of ice crystals.

of their great height they appear to move slowly even though they may be travelling a hundred miles an hour or more. Cirrus clouds indicate the presence high aloft of air whose moisture content (and therefore whose origin) is different from the air immediately beneath the cloud. They form in regions where there are unusually strong updrafts to carry water vapour so high and are usually the first sign of an approaching low-pressure system, perhaps 500 or 600 miles in advance of the front.

But the cirrus are not the most remote clouds that veil the upper reaches of earth's aura. Mother-of-pearl clouds spread their iridescent film at thirteen to eighteen miles high – far into the stratosphere. They are especially beautiful just before sunset and they can still be seen as long as three hours afterwards. Like the cirrus, they are composed of ice crystals and are brilliantly banded with colour. They are most often observed in extreme northern regions, Alaska and Scandinavia, when the air is abnormally clear. Sigrid Undset described a sighting of mother-of-pearl clouds in these words:

> The northern silhouette . . . was extraordinarily clear. Against the golden center the cloud changed colour in an infinite variety of red-violet, lilac-pink, red and orange tones. Up towards the zenith and downwards in a southerly direction its contours appeared as if blown away by the wind – and surrounded by a fan-shaped, turquoise green, gold-rimmed field it slowly vanished into the air . . .[12]

In a general way it is known that mother-of-pearl clouds occur at the meeting place of waves in the stratosphere. But the exact conditions that cause these rarefied currents to give birth to ice crystals have not been identified.

On 19 May 1910, the day the earth passed through the tail of Halley's comet, an exceptionally lovely display of mother-of-pearl clouds was seen. Perhaps the dust of the comet's tail caused the supercooled air of the lower stratosphere to precipitate into a shower of ice crystals just as cloud-seeding triggers condensation in the lower atmosphere.

The hypothesis of a seeding mechanism seems to be further substantiated by measurements made of volcanic dust clouds. In October 1974, the volcano El Fuego, near Antigua, Guatemala, erupted with great force, showering rocks on near-by villages and depositing ashes as far north as Mexico. The dust cloud thrown up from this eruption was tracked by NASA physicists with a new, very accurate system known as lidar. Pulses of red lasar light beamed at the dust cloud were reflected back to earth and the time for this round trip was very precisely measured. Since the speed of light is known, the distance to the cloud could thus be determined. It was found that shortly after the eruption there were two layers of clouds at 9·6 and 12·2 miles. Months later the clouds had joined into a single one with maximum density at about 11·5 miles but extending up to 15 miles, within the range of altitudes at which mother-of-pearl clouds have been sighted.[14] Volcanic ash remains in the atmosphere for several years and may be one of nature's ways of triggering the crystallization of these shimmering veils of ice.

The loftiest clouds of all are known as noctilucent clouds because they are lighted by the sun almost all night. They are rather rare occurrences and have been seen only during the summer months at high latitudes in both the Northern and the Southern Hemisphere. About a quarter of an hour after sunset these mysterious clouds begin to materialize in a sky that had earlier seemed to be perfectly clear, between the twilight arch at the horizon where the sun has set and the darkening blue vault of the sky overhead. By an hour after sunset they show up very clearly, reflecting the sunlight with a silvery-blue glow which shades into golden yellow near the horizon. Their delicate feathery ripples of foam look like waves in a phantom sea.

The altitudes and speed of movement of these clouds have been quite precisely measured. They always occur at heights of approximately 50 miles, above 99·9 per cent of the atmosphere and at least 38 miles above the troposphere, the region where clouds normally exist. This is the location of the mesopause, the thin layer of extremely cold atmosphere that separates the mesosphere and the ionosphere. The mesosphere is a relatively quiescent region where few energy-releasing reactions occur. It reaches its coldest point during the summer, when it falls to $-225°F$ $(-143°C)$. It is in this layer and at this time of year that the noctilucent clouds are most often sighted.[15] Such very low temperatures, of course, would cause condensation of any water vapour that exists there, but measurements of the size and reflectivity of the clouds indicate that dust as well as ice is present. What is the source of this dust? Does it come from the earth or is it brought in by shooting stars and comets from outer space? In 1908 an enormous meteorite estimated at 1,000 tons fell in Siberia. This event was immediately followed by the appearance of very striking noctilucent clouds.[16]

In 1962 some of the questions about noctilucent clouds were answered. An experiment was conducted in Sweden by a team of American scientists working in cooperation with experts from the Institute of Meteorology at the University of Stockholm. The equipment, provided by the United

States National Aeronautics and Space Administration, consisted of four rockets capable of lofting a 90-pound load to a height of 75 miles. Each rocket carried a device for collecting a sample of the cloud particles, a parachute for returning the sample to earth, and a radio beacon to direct the researchers to its landing site. During the late summer of 1962 experienced observers throughout Scandinavia watched for these characteristic cloud formations. Airline pilots also cooperated and the whole population of Sweden was alerted by press publicity to report clouds that were visible long after sunset. The first sighting was made in early August and a rocket was fired into the cloud. The other three rockets followed later that month. The particles collected and returned to earth proved to be tiny grains of dust which had apparently been coated with ice. The size and the chemistry of the particles indicated an extraterrestrial origin. They were considerably larger than the maximum size of terrestrial dust or volcanic ash that could have been carried to these heights. The dust within the clouds themselves was much more densely distributed than in the cloudless space at the same altitudes. The participating scientists felt confident that the particles came from outer space, but they could not explain the uneven distribution of this dust. Could it be caused by chance alone or is there some unidentified force that gathers in and concentrates a wisp of tenuous matter from a comet's tail, fragments from a shooting star, or a filament of cosmic dust swept up as the planet wheels majestically in its elliptical orbit around the sun?[17] These diaphanous swirls of matter from other worlds are dense enough to make their presence known to our relatively clumsy senses. But we see only the crest of the waves, the accumulations of matter thick enough to reflect the sun's light. We are unaware of the great ocean of atoms that rains onto our world from outer space.

In just the same way we are unable to sense the countless billions of water vapour molecules floating in unsaturated air until there are enough of them joined together to catch and turn back the light. Clouds are the spray thrown up by the clashing currents of the atmosphere, the spindrift of waves breaking in an otherwise invisible sea.

CHAPTER FOUR

CYCLES IN THE BIOSPHERE

Here in my curving hands I cup
This quiet dust . . .
A bit of God himself I keep
Between two vigils fallen asleep.

    – John Hall Wheelock[1]

In March 1775, just six weeks before the first shots of the American Revolution were fired at Lexington and Concord, an unusual scene was being enacted in a firelit room in an imposing castle near Calne in England. A slender aristocratic-looking gentleman wearing a dark frock coat and a white clerical collar picked up a tiny mouse by the nape of its neck and deftly transferred it from its cage to a glass bell jar that was inverted over a pan of water. A wooden platform floated on the water. The gentleman placed the mouse on the platform and removed another mouse from its cage, placing it in a second bell jar. Then he glanced at his pocket watch and, seating himself comfortably, began to play his flute while he kept a watchful eye on his curious experiment.

Joseph Priestley was a very unconventional clergyman. Not only did he have interests that involved him in queer scientific pursuits, but he also held political and religious views that were radical for his day. He believed in human liberty and equality and did not believe in original sin or eternal damnation. A wealthy English nobleman, William Petty, Earl of Shelburne, shared some of Priestley's views and admired his curiosity about natural phenomena. For several years now he had acted as the clergyman's patron, providing him with an annuity, a summer residence at Calne, and a winter home in London. Aside from his duties as librarian and companion to Lord Shelburne, Priestley had a great deal of time to devote to his hobbies.[2]

Just six months earlier Priestley had discovered how to prepare an interesting new gaseous element by burning a form of mercury oxide known in those days as mercury calx. He did not realize that the element he had isolated was oxygen. He only knew that it was a remarkable gas that caused candles to burn with a very vivid flame and charcoal to blaze much more brilliantly than in air. In several ways the newly isolated element seemed to be better than air. This was a revolutionary idea in itself because at that time most people believed air to be a pure, simple, elementary substance. But Priestley's experiments had convinced him that there were differences in air depending on its history. There was good air and there was noxious air. Priestley thought that perhaps his new gas might be even more wholesome than the air we usually breathe. He had discovered that a small animal, such as a mouse, placed under a bell jar which contained ordinary air would survive for only a short time. Something undesirable happened to the air when it was breathed by animals. He had also discovered that if he put a small plant, such as a sprig of mint, into the air which had been rendered unfit by the fact that a mouse had breathed it and finally died in it, the plant would begin to sprout and grow prodigiously – much faster than it would grow in ordinary air.

'In no other circumstances,' Priestley said,

> have I ever seen vegetation so vigorous as in this kind of air, which is immediately fatal to animal life. . . . This observation led me to conclude that plants, instead of affecting the air in the same manner with animal respiration, reverse the effects of breathing, and tend to keep the atmosphere sweet and wholesome, when it is become noxious, in consequence of animals either living and breathing or dying and putrefying in it.[3]

On this March day Priestley was testing the characteristics of the new element which he had prepared from mercury calx. The evening before he had set out in the castle mousetraps that were especially designed to allow him to remove the animals unharmed. Now he chose two of the most vigorous of these mice, placed one in the bell jar containing ordinary air and one in the jar containing two ounce-measures of the new gas. 'Had it [the new gas] been common air,' Priestley observed,

> a full-grown mouse as this was, would have lived in it about a quarter of an hour. In this air, however, my mouse lived a full half-hour; and though it was taken out seemingly dead, it appeared to have been only

exceedingly chilled; for, upon being held to the fire, it presently revived and appeared not to have received any harm from the experiment.[4]

In the other bell jar, as Priestley expected, the mouse incarcerated in ordinary air became unconscious after fifteen minutes and could not be revived.

Joseph Priestley had found a method for preparing a sample of pure oxygen and, although he failed to identify this gas as a simple chemical element, he received the credit for the discovery of oxygen. A Frenchman, Antoine Lavoisier, repeating Priestley's experiment, made the proper identification about four years later, and launched the science of modern chemistry.

Some of the special characteristics of oxygen were revealed by the bell-jar experiments. Oxygen is essential for animal life and it supports the burning process more efficiently than the same quantity of ordinary air. Priestley's experiment with the mouse and the mint also provided some very important clues about the nature of the atmosphere. A mouse confined alone in a small quantity of air very quickly sickens and dies. But when a plant is introduced into this toxic air and allowed to grow there for several days, the air is restored. A mouse placed back in it lives as long as it might live in an equal quantity of fresh air.

Again Priestley did not draw all the right conclusions from this experiment. The science of chemistry was in its infancy and did not provide him with the proper concepts for making such an analysis. We know today that plants need carbon dioxide, which they take in through their leaves, and, using the energy of sunlight, they combine the carbon dioxide with water to synthesize organic molecules such as sugar and starch which act as storehouses of energy. Oxygen is left over in this process and is rejected by the plant. On the other hand, the mouse needs oxygen, which it breathes in and uses in the process of breaking down the organic molecules of its food. This reaction produces energy which enables the mouse to move and breathe and go on living. The reaction also yields carbon dioxide as a waste product; so this gas is breathed out by the animal and enters the air again. Thus the mouse and the mint each produce the conditions that are favourable for the other.

Priestley's experiment is an elegant demonstration of the way natural systems work together to produce a balanced economy of use and re-use.

Chemicals that are waste products for one part of the system serve as raw materials for another part. They are all recycled and there is no net waste in the system. Marvellously balanced chemical cycles of this kind are going on every day around the whole surface of the planet. However, when we move out of the laboratory into the real world the relationships become considerably more complex.

On and near the surface of the earth there are four spheres of matter that interact and form an integrated whole: the atmosphere, the water, the crust of the earth, and the mass of living matter. The elements that are found in the air all circulate through these four realms and no one of the realms can be understood without considering the action and influence of the other three. The name *biosphere* has been coined to designate the envelope of air, water, and soil in which life exists.

In the biosphere the most common constituent is oxygen. It accounts for one out of three atoms in every molecule of water in the oceans, rivers, and clouds. It is present in combination with calcium, silicon, and many other elements in the sedimentary rocks and sands of the earth's crust. It is one of the fundamental constituents of living matter, comprising about one fourth of all the atoms in vital molecules. And as free oxygen it makes up about 21 per cent of the total volume of the air. Some of the other important components of the atmosphere, such as carbon dioxide, are compounds of oxygen.

In its free form as a gas oxygen is believed to be a comparative new-comer to the biosphere, building up in appreciable quantities only after the invention of photosynthesis. Before that time the world must have been unbelievably different from the one we know today. It was relatively colourless – all blue and grey except for the transient hues of the sunset and the rainbow. There were no flowers, no brightly plumaged birds, no richly variegated mantle of green to cover the naked rocks of the continents. The wind and the rain, the pounding of waves and the rumble of thunder were the only sounds that broke the stillness of that pallid scene where no life stirred, no spontaneous movement ruffled the surface of the sea. The first very simple forms of life which evolved in water must have lived and multiplied far below the surface, hiding in that dim world from the ultra-violet sunlight which beat down pitilessly on the barren land and the sterile layer of surface water. These rays are so energetic that they would have damaged the delicate molecules of living matter and even broken up the organic molecules floating near the top. The earliest life forms must, therefore, have lived in the half-lighted depths so that water could act as a shield against the ultra-violet rays. (It has been estimated that about thirty feet of water was required to provide adequate protection.)[5]

These tiny self-replicating creatures were unable to utilize the energy of sunlight to maintain their own life and movement. They probably subsisted on the supply of organic molecules which had accumulated in the sea and, as this supply became depleted, their numbers were limited. Slowly, over millions of years, they mutated into more complex forms and then one day a cell appeared which contained a new form of molecule – chlorophyll. The first true plant had been created, an event almost as important as the appearance of the first living organism. This cell could manufacture its own food using sunlight, an ability which gave it an enormous advantage over its neighbours. Fossil remains of very primitive photosynthesizing organisms bear witness to the fact that this critical event in the history of the planet must have taken place at least three billion years ago.

Biologists believe that they can trace the gradual adjustment of life to the changing conditions that were brought about by the new chemical process that had come into the world. Algae similar to the blue-green algae of today were probably the first organisms to contain chlorophyll. They lived at the bottom of shallow lakes and ponds where the water was just deep enough to screen out the ultra-violet and not too deep to prevent the visible light from penetrating. As the colonies grew, the level of oxygen began to build up in the atmosphere, and ozone (the three-atom molecule of oxygen) also began to accumulate. When the concentration of oxygen reached about 1 per cent of the present level, perhaps 600 million years ago, ozone shielded out enough of the ultra-violet to allow cells to live in the upper sunlit layers of the lakes and seas. Photosynthesis could take place much more effectively there and the rate at which free oxygen escaped into the atmosphere increased geometrically.

But free oxygen was both a friend and a foe to early life forms on earth. Molecular oxygen is extremely reactive, combining spontaneously with organic compounds. In the metabolism of sugar and starch molecules, for example, the reaction with oxygen releases large amounts of energy which can be used by the organism. On the other hand, this reactivity makes oxygen so potentially dangerous to fragile living tissues that all organisms which used it had to develop chemical ways of shielding themselves from its destructive nature. Many of the simple one-celled creatures that existed in the primordial seas would have been killed by direct exposure to an atmosphere containing oxygen. (Louis Pasteur discovered that certain very sensitive microscopic organisms cannot tolerate oxygen concentrations above 1 per cent of the present level.)[6] Some of the primitive organisms continued to exist in old ways that protected them from contact with this dangerous new element. They stayed deep under water, buried themselves in mud, or hid themselves under layers of sediment. Their descendants – the anaerobic bacteria, yeats, fungi – are still plentiful today, exploiting this ancient way of life.

Other organisms found a more creative solution. Through the gradual process of mutation and evolution they developed the protective mechanisms that enabled them to take advantage of the opportunity and to use free oxygen safely, a feat that was achieved with the help of *enzymes.*

Enzyme is the name given to any *catalyst* that occurs in living organisms. A catalyst encourages and accelerates one specific chemical reaction without itself being used up in the process. Even though the catalyst may undergo change while the reaction is occurring, it is reformed at the end and is ready to perform the same expediting function for an endless stream of identical reactions, providing a shortcut along which the process can be led again and again. All living organisms make use of this remarkable type of compound to control and direct the complicated reactions that take place at the cellular level, and many processes that seem to be pure magic turn out to be regulated by enzymes.

Photosynthesis itself is governed by a packet of enzymes working together with molecules of the green pigment chlorophyll. The chlorophyll molecules act like little antennae reaching out to absorb energy from sunlight. A small bundle of this energy is transmitted from one chlorophyll molecule to another until it reaches a location where a reaction controlled by enzymes can occur and the energy causes the ejection of a high-speed electron from the molecule. This electron can be made to do chemical work and thus light is converted into chemical energy.

In a similar manner the enzymes that direct the reactions that occur when molecular oxygen enters a living cell encourage the reactions that make the energy available from oxygen useful to the organism or release its energy in ways that are not injurious. All cells of higher organisms contain such enzymes and the evolution of this regulating system was probably an essential step in the development of the more complex forms of life after oxygen became an important component of the atmosphere.

About 400 million years ago, when the concentration of oxygen had risen to approximately 10 per cent of the present level, the ozone layer began to provide sufficient protection to allow life to climb out of the sheltering seas and colonize the rocky shores. Ever so slowly, but as irresistibly as a tide, the flood of green vegetation flowed across the land.

For the next 300 million years green was the only colour added by plant life to the planet earth. Mosses and ferns, pine and spruce forests, giant redwood and sequoia trees dominated the landscape. But grasses did not exist and there wasn't a flower or blooming plant to be found anywhere on earth even as late as the beginning of the Cretaceous period, when dinosaurs roamed the land and pterodactyls soared on huge leathery wings over the seas.

Then about 100 million years ago another important innovation occurred. Angiosperms, the true seed-bearing plants, made their appearance, and in a brief instant of geological time, they had spread around the planet. Each seed of the angiosperm is a fully equipped embryonic plant, carrying its own store of food and able to travel on the winds far from its parent plant. These new life forms had also developed a new method of reproduction which did not rely on the vagaries of the wind for pollination but employed a display of attractive colours and fragrant messages sent on the breeze, enticing insects to pollinate the plant. Along with the flowers evolved honeybees and ruby-throated hummingbirds and bright butterflies like blossoms released from their stems and floating free on the wind. Grasses carpeted the bare soil where the forests had not taken hold; water lilies opened their pink-and-white starlike blooms; and buttercups turned whole hillsides into drifts of gold.

The earth began to look as various and beautiful as it does today. But most important of all, the angiosperms produced grains and fruits that provided the concentrated food needed for the warm-blooded active animals that were beginning to appear on earth. They set the stage for the evolution of mammals and the emergence of man. As Loren Eiseley says:

'In that moment the golden towers of man, his swarming millions, his turning wheels, the vast learning of his packed libraries, would glimmer dimly there in the ancestor of wheat, a few seeds held in a muddy hand.'[7]

This surge of multicellular life across the surface of the planet was all derived from that first plant cell that was able to capture a packet of energy from light and release it later to energize the dynamic process of living. Not only vegetation but also all animal life depends upon chlorophyll. Animals – who never learned how to utilize energy directly from sunlight – developed a parasitic way of life, consuming the energy-packed molecules synthesized by the plants and using them to produce the continual flow of energy needed for growth and movement.

The miracle of plant growth is so familiar to us that we do not stop to wonder how large solid objects can materialize out of what seems to be thin air – or at least air supplemented only with a little water, sunshine, and soil that does not appear to be used up. Every year by this process hundreds of billions of tons of starch, sugar, and protein molecules are built up, providing the food for the entire life-support system on earth. Approximately half the dry weight of all this vegetation is composed of carbon derived from carbon dioxide in the atmosphere.

Carbon is the basic building block of all organic molecules. Its atom acts like a neatly symmetrical tetrahedron that can link up with other atoms in an almost infinite variety of forms. In order to synthesize these carbon compounds, energy is required, and the energy that is stored in the compound can be released again by the various processes, such as burning and metabolism, that take place when oxygen is present. All animals breathe out the carbon dioxide that is left over from the metabolism of their food. Plants also must use some of the food that they themselves synthesize and in this process carbon dioxide is respired. Even the earth itself breathes out carbon dioxide; large amounts are emitted in every volcanic eruption. The leftover products of digestion and dead organic material of all kinds are reconverted by fermentation and decay into simple chemical compounds. This transformation is achieved by bacteria and fungi with the help of enzymes, and one of the results is the re-formation of carbon dioxide, which returns to the atmosphere.

This natural carbon cycle maintains the very small component of carbon dioxide in the air – approximately one third of 1 per cent – a low concentration that is favourable to animal life. Experiments have shown that concentrations of carbon dioxide over 1 per cent begin to be toxic for laboratory animals as well as for human beings. At 5 or 6 per cent the rate of respiration increases; at 8 per cent vertigo and dizziness occur; 9 per cent can be fatal. Above this level a candle stops burning and an animal quickly expires.[8] Plants, on the other hand, can tolerate much higher levels of carbon dioxide. In fact, as Priestley discovered with his mint, plants grow more rapidly and luxuriantly in the presence of increased levels of carbon dioxide (if adequate supplies of other ingredients such as water are present). This fact suggests that plants may have evolved in an atmosphere that contained more of this gas than our atmosphere does today. The proportion may have changed, fluctuating as ice ages came and went and gradually declining as plant life became a more important factor in the biosphere.

At the present time we can observe some minor fluctuations in concentration due to the variations in vegetation with the time of year. In the spring, when there is a great burgeoning of growth and the mantle of green spreads rapidly poleward across one entire hemisphere, the consumption of carbon dioxide exceeds the return. From April through September the atmosphere north of 30° north latitude loses about 4 billion tons of its carbon dioxide content. But, large as this effect might appear to be, it reduces the carbon dioxide concentration in the air only by about 3 per cent.[9]

Occasionally it happens that dead, decaying material is trapped under a layer of earth or sediment and is prevented from completing the breakdown portion of the natural carbon cycle. The trapped organic materials, gradually compacted and heated over millions of years, form coal, oil, and gas deposits which are storehouses of energy-packed carbon compounds.

A carbon cycle similar to that on land takes place in the sea. Carbon dioxide is readily soluble in water. The amount dissolved in the sea depends on the concentration present in the atmosphere and on the temperature of the water. Small primitive plants such as algae use the process of photosynthesis to convert the dissolved carbon dioxide into organic molecules, releasing oxygen just as the plants on land do. A portion of this oxygen remains in the sea in a dissolved state. The colder the water, the more gases like oxygen and carbon dioxide it can contain in

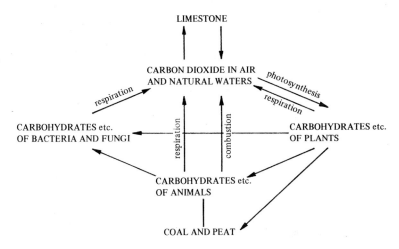

Fig. 2. Carbon cycles on land and in the sea.

solution. As the water warms, the gases return to the atmosphere. The vast number of minute photosynthetic organisms in the ocean provide food for the larger forms of marine life. When these organisms die, bacteria and fungi break down the organic matter, converting it to carbon dioxide and water.

Just as on land, a small fraction of the carbon compounds synthesized by living matter in the sea escapes decay and fermentation – tiny particles of lime and silica, shells and bones, remain intact and fall slowly downward. Gradually, over aeons of time, the skeletons of untold numbers of marine organisms form great drifts on the ocean floor. There they are compacted by water pressure and by the other sediments that continue to build up over them. Sometimes, by the accident of a major tectonic change in the earth's crust, these deposits are lifted high above the depths where they were laid down. Such is the origin of the white cliffs of Dover, which are almost pure calcium carbonate, composed of trillions of skeletons of microscopic zoo-plankton. Here in these cliffs lies an enormous storehouse of carbon in a soft-textured white form that is in startling contrast to the jet-black stores of carbon that lie in coal seams around the world. Each of the tiny creatures whose shells are compacted here and the delicate plants whose fossil shapes are imprinted in the deposits of coal, lived for only a few weeks or months, but their forms have been preserved for as long as three hundred million years. For a brief moment of life they absorbed their minute allotment of solar energy, performed for an instant the magic of transforming it into movement and growth, and wove their own highly complex and individual forms out of the basic components of living matter. Thus they entered for all time into the vast cyclic movements that circulate a few essential chemicals from air and water to living tissue and then back again to the wind and the sea.

Hydrogen, carbon, oxygen, and nitrogen – these four elements make up most of the air and almost all of living matter. Nitrogen is by far the largest component of the earth's atmosphere – 78 per cent is elementary gaseous nitrogen. On the other hand, the amount of nitrogen in living matter is comparatively small – less than 1 per cent. Given these facts, it is an interesting paradox that the food supply of organisms on earth is more limited by the availability of nitrogen than any other of these nutrients. Most of the great ocean of nitrogen in the atmosphere is in a form that cannot be used directly by multicellular plants or animals. Nitrogen gas, which has two atoms in each molecule, is an extremely stable chemical; it does not readily enter into chemical combinations with other elements. Before reactions can occur the nitrogen molecule must be broken down into single nitrogen atoms, which then do enter into a great variety of chemical reactions. A large amount of energy is required to dissociate the nitrogen molecule so that the individual atoms can enter into other combinations and some of this energy is stored in the resulting compound. The process is known as 'fixing' nitrogen.

Scientists theorize that nitrogen was probably available in a more usable form on the primitive earth. Ammonia (a compound of nitrogen and hydrogen) is thought to have been one of the principal constituents of the primordial atmosphere. This compound can be very easily assimilated by plants and used to build living tissue. But when free oxygen began to appear it reacted with the ammonia, forming water and releasing the gaseous nitrogen which is such a predominant component of the atmosphere today.

Supplies of fixed nitrogen are now somewhat limited in the biosphere. One of the best natural sources is lightning. In the searing heat of the electrical discharge the nitrogen molecules are torn apart and the single nitrogen atoms then combine with oxygen to form nitrogen oxides. These in turn react with water to form nitrates. In this way a thunder and lightning storm delivers a large amount of fixed nitrogen to the air and soil.

Lightning is a very obvious source for the large amounts of energy demanded for nitrogen fixation. But the natural process that accounts for the greatest share of fixed nitrogen is accomplished in silence and usually in darkness by some of the smallest and most primitive of organisms. Certain types of anaerobic bacteria that cannot tolerate oxygen live underground in colonies near or actually on the roots of a number of larger plants – clovers, alfalfa, peas and beans, cycads and ginkgo trees. These colonies of single-celled organisms cooperate with the plants in a symbiotic relationship. Some types are parasitic, depending on the plants for their energy; others obtain their energy directly from sunlight. All of these colonies provide their host plants with nitrogen in a form that can be used to build protein molecules. This trick is accomplished with the aid of an enzyme called *nitrogenase*.

The details of this enzyme-mediated transformation are not completely understood, but the remarkable nature of the process can be appreciated when we consider that the only other important natural process for fixing nitrogen involves the power of a lightning stroke, a phenomenon that heats the air up to 18,000°F (10,000°C) in a fraction of a second.

The most efficient method men have invented for fixing nitrogen involves the use of a catalyst in combination with hundreds of degrees of temperature and thousands of pounds of pressure. This same transformation is accomplished at ordinary temperature and pressure by some of the smallest living organisms (25,000 of these bacteria laid side by side would measure scarcely a half inch). And they do their work so efficiently that although it provides the largest source of naturally fixed nitrogen, the total amount of nitrogenase in the world is probably less than ten pounds.[10]

After nitrogen has served its purpose in living matter, another army of bacteria reduces the protein molecules of dead plants and animals to simpler compounds – nitrates and nitrites – that can be used again by plants. And finally, a third group of soil inhabitants, called the 'denitrifying bacteria', breaks down the nitrates and nitrites by catalytic reactions, liberating gaseous nitrogen, which returns to the air.

Thus the passage of the nitrogen atom between the realm of living matter and the realm of the atmosphere is expedited by enzymes which ferry each molecule across the interface of these two realms, like Charon ferrying souls from earth to Hades across the river Styx. Carbon atoms also and even the treacherous oxygen atoms are taken in charge by similar ferrymen. And although biochemists can chart the course of these passages, they have not been able to whistle up the ferrymen themselves.

There is much that is still unknown about the way in which nature performs these miracles of balance and how it responds to large changes that might disturb the equilibrium. For example, when more fixed nitrogen is added to the soil (from fertilizers and air pollutants washed out by rain), are the colonies of denitrifying bacteria able to keep pace and complete the liberation of gaseous nitrogen back into the atmosphere? We really do not know, and these gaps in our knowledge are particularly disturbing in view of the fact that human activities are making very significant changes in the quantities of the chemicals present in the soil and the air.

But in the absence of any interference from man these wonderful natural processes connect the four portions of the biosphere, recycling all wastes and maintaining the precise combination of chemicals ideally suited to the support of the living matter that exists on earth today.

Reading these facts, we might be overcome with a sense of our remarkable good fortune. Isn't it lucky, we might say, that our atmosphere contains the principal ingredients necessary for life and that these are present in the most benign proportions? Isn't it fortunate that the very wavelengths of light necessary for photosynthesis are able to pass through our atmosphere while those that are injurious are screened out? But actually these remarkable facts need not be ascribed to luck at all. The atmosphere that envelops our planet today developed slowly in response to the changing nature of the land, the rise and rearrangement of the earth's waters, the growing mass of living organisms. Over many millennia life evolved to make maximum use of the conditions it found. As new elements appeared it invented new ways to control and take advantage of them.

Our environment is still changing; the crust of the earth continues to move; the waters flow in ever shifting channels; and the composition of the atmosphere is changing as time goes by. Eventually life would undoubtedly adjust to these altered conditions, but the time scale for nature to make such adjustments by the slow process of evolution is measured in millions of years.

In the sweep of cosmic processes nature is indifferent to the fate of the individual, as witnessed by the tiny zooplankton whose finely shaped shell has lain compacted in the white cliffs of Dover for millennia, while continents have drifted apart, mountains have been raised up, and glaciers have advanced dozens of times across the face of the earth. For us, a century – perhaps a decade or two – is all. We must look out for ourselves if we care about the changes that occur on our planet within these short time spans – if we care about ourselves and the other ephemeral beings that share with us this brief instant of earth-time, the humming bird, and the ginkgo tree, and the fern in the forest.

CHAPTER FIVE

MAN'S AURA

. . . this most excellent canopy, the air, look you, this brave o'erhanging firmament, this majestical roof fretted with golden fire, why, it appears no other thing to me than a foul and pestilent congregation of vapours.

– William Shakespeare[1]

In St James's Park the bright October afternoon was filled to over-flowing with a glow of warm sunshine and the flutter of pigeon wings. Many Londoners were out enjoying the day, strolling along the paths, scuffing the leaves that had settled on the walks and the well-kept lawns like a scattering of gold coins. Flocks of ducks and geese rode the quiet waters of the pond and along its shores rose bushes were heavy with late bloom. On the grass beside the pond a little old woman was feeding the pigeons bread crumbs from a brown paper sack. They clustered around her, perching on her shoulders and pecking the crumbs out of her hands. Nearby an elderly man sat on a park bench reading the *Evening News*. He held on his lap an empty bag in which he, too, had brought an offering for the birds – as he did every afternoon, weather permitting. Leafing through the pages of his paper, his eye was caught by a feature article, 'The Menace of Smog':

> Stand on the heights of Hampstead or Blackheath . . . and it will be seen that the grey dome of St Paul's stands above a layer of mist stretching from the cranes and derricks of the Pool to the gasometers of West London. That layer is smog, an urban combination of smoke and fog. It is not only smoke particles, but also an almost infinite variety of chemical compounds, in solid, liquid and gaseous form. In that air are varying amounts of soot, ash, carbon monoxide, sulphur dioxide, oxides of nitrogen, hydro-carbons, organic acids, methane, acetylene, phenols, ketones, ammonia, alcohols and much else besides – a veritable chemical storehouse . . . Smog is always with us, as we shall probably learn this month.[2]

The old man snorted. Smog! Reporters must be hard up for news, trying to get people all stirred up about anything as common as fog. Sure, London had bad fogs sometimes – real pea-soupers. It had always been so as long as he could remember. He glanced up at the bright scene before him. The water sparkled where a gull dipped and skimmed over the surface. A swan climbed out onto the bank and preened its feathers. Beyond the far side of the lake, Whitehall with its domes and turrets rose shimmering white and grey. Just faintly softened by haze, it looked like the dream palace of some Far Eastern potentate. With days like this, thought the old man, Londoners could put up with a little fog occasionally.

The old man did not know that for hundreds of years some prophetic reporters and writers had been trying to warn the British public about the potential danger of smoke pollution in London. In 1661 a diarist named John Evelyn circulated a pamphlet he had written on 'the smoke of London'. Addressing himself to King Charles II, he wrote:

Sir, it was one day when I was walking in Your Majesty's Palace at Whitehall (where I have sometimes the honour to refresh myself with the sight of Your Illustrious Presence, which is the joy of Your People's hearts) that a presumptuous smoke issuing from near Northumberlandhouse, and not far from Scotland-yard, did so invade the Court, that all the rooms, galleries and places about it, were filled and infested with it, and that to such a degree as men could hardly discern one another from the cloud, and none could support, without manifest inconvenience. It was this alone and the trouble that it needs must [give] to Your Sacred Majesty, as well as the hazard to Your Health, which kindled this indignation of mine. Nor must I forget that Illustrious and Divine Princess, Your Majesty's only Sister, the now Duchess of Orleans, who, late being in this city, did in my hearing, complain of the effects of this smoke both in her breast and lungs, whilst she was in Your Majesty's Palace . . .

[Londoners] breathe nothing but an impure and thick mist, accompanied by a fuliginous and filthy vapour, corrupting the lungs, so that catarrhs, coughs and consumptions rage more in this one city, than in the whole Earth . . .

That this glorious and ancient city, which commands the proud ocean to the Indies, and reaches the farthest Antipodes, should wrap her stately head in clouds of smoke and sulphur, so full of stink and darkness, I deplore . . . For when in all other places the air is most serene and pure, it is here eclipsed with such a cloud of sulphur, as the sun itself, which gives day to all the world, is hardly able to penetrate, and the weary traveller, at many miles distance, sooner smells, than sees the city to which he repairs.

This is that pernicious smoke which sullies all of [London's] glory, superinducing a sooty crust or fur upon all that it [touches], spoiling the movables, tarnishing the plate, gildings and furniture, and corroding the iron-bars and hardest stones.[3]

In spite of the persuasive words of John Evelyn, King Charles did not take any action to correct the pollution in his palace or in his city. For almost three hundred years thereafter similar warnings were sounded and went unheeded. Aside from a few very ineffectual attempts at smoke control during the nineteenth century, nothing was done to control the increasing air pollution in London. Now in 1952 there was still no official recognition that a serious problem existed – no sense of impending disaster.

During the next six weeks the weather turned sharper. There were beginning to be traces of frost in the mornings but the days were reasonably pleasant. People still came to the park. Wednesday, 3 December, was a beautiful day with a blue sky filled with feathery cumulus clouds. Young women pushed their babies in strollers; children and old people brought crumbs and seeds for the birds. But December days are short in England and as the chill of evening began to come on, the park emptied very quickly.

Thursday was not as pleasant as Wednesday. The air was dank and chilly; a haze obscured the sun. Still there were a few habitués in St James's Park, taking a brisk walk or sitting huddled on park benches reading the *Evening News*. The paper mentioned that considerable fog might be expected during the next twenty-four hours.

On Friday, Londoners woke to find that the first pea-souper of the winter had indeed arrived. The fog was so thick that they could hardly see across the street. The cold humid air seemed to be penetrating even into the buildings. Millions of people started up coal fires in their grates to take off the chill; they turned on their ranges and their toasters to make breakfast and the power plants at Battersea and Fulham poured on more fuel to meet the early-morning demand. The smoke from all these fires did not rise in tall plumes as usual. It oozed out horizontally and drifted down into the mass of chilly stagnant air which had moved over London. This air hung so close to the ground that on the hilltops like Hampstead Heath and Shooter's Hill, at elevations of only 400 feet or so, a bright sun still shone and people could look out over the top of the sea of dark yellow fog that lay like a smothering blanket across the city.

All day Friday, Saturday, and Sunday the fog grew denser. Traffic ground almost to a standstill; bus service was severely curtailed. The radio and newspapers mentioned these inconveniences, but they did not warn that there was any health hazard. People who were out in the smog, however, noticed a burning in their throats and a tightness in their chests. Their breathing became laboured; and many vomited. Even young people exposed to the pollution for more than a few hours began coughing painfully, bringing up phlegm, black as soot.

One man who had been riding a bicycle recounted a strange experience. In the unearthly quiet of the foggy street he heard a desperate honking sound coming towards him. He dismounted from his bicycle and stood his ground, prepared for any emergency. Suddenly a swirl of white broke the fogbank before him and an enormous swan with wings flapping frantically lumbered into sight, its long neck stretched awkwardly forwards and its beak wide open, emitting the wild trumpeting sound. The bicycle rider knew that the place where he encountered the swan was at least half a mile from any water. The swan must have climbed out on the bank, taken the wrong turning, and got lost in the fog.

Many of the elderly men and women who had gone regularly to the parks with their bags of crumbs for the geese, the swans, and the pigeons were also lost in the fog. The old, the lonely, the sick, and the indigent represented the largest proportion of the victims of this disaster, which held London in its grip for four days and nights. After the pollution cleared people began coming into their local police stations to report missing neighbours:

'Old Mrs M—, who lives alone in the building next door to me, hasn't been seen for days. She always used to take a walk out in the afternoons.'

'Mr P—, who has a furnished room above my store, hasn't stopped in for his groceries or his evening paper since Friday.'

The police investigating these stories found many bodies of people who had died alone when the smog penetrated the flimsy buildings in which they lived. Several days after the fog had cleared a unit of the Metropolitan Police picked up a body that was floating in the Thames. The officer who accompanied the body to the morgue was shocked at the scene there. It was crowded with corpses, lying row upon row, their uncovered faces staring at the ceiling. Why were the corpses not covered? he asked. There is nothing to cover them with, an attendant replied. In the last few days so many people had died that the Borough of Southwark had run out of shrouds.[4]

Four thousand people died after a brief (sometimes only a few hours') exposure. Those in marginal health succumbed first, but many active, healthy people in the prime of life were also victims of the smog. They suffered damage to their lungs and bronchi that left them semi-invalids for the rest of their lives.

'In the past hundred years,' reported a medical journal, 'only the peak week of the influenza pandemic in November 1918 produced more deaths over the expected normal than did the man-made fog . . . Even the cholera epidemic of 1866 could not quite equal it.'[5]

On Sunday afternoon the air in London, according to official estimates, had contained three to twelve times the normal amount of sulphur dioxide. But these estimates may have been considerably below the true figures, for the pollution was so heavy that the instruments were not working properly. At the Ministry of Works the filters in the air-conditioning system clogged up with particulate matter and had to be changed fifty-four times more often than normal.

All the smoke and other products of combustion from the fires that heated the homes and ran the industry in London were trapped in a shallow layer of cold air which lay over the city. For four days and nights the poisonous effluents continued to collect in it like raw sewage flowing into a cesspool.

The atmospheric condition that allows this nightmare to occur is known as a temperature inversion. The air next to the ground is colder than the air aloft, and this reversal of the usual situation (where the temperature falls about $3.6°F$ or $2°C$ every thousand feet of altitude) suppresses the normal vertical flow of air that carries pollutants high into the atmosphere, speeding their diffusion. Temperature inversions occur very frequently at night throughout the temperate zones – at least two out of every five nights throughout most of the United States and Europe.

In fair weather when the sky is clear, the ground loses heat very rapidly after sundown. The cool surface reduces the temperature of the atmosphere near ground level, forming a low-lying layer of colder air. In most cases this condition is dissipated within a few hours after dawn when the sun's radiation has warmed the surface of the earth again.

Sometimes, however, the inversion persists because other factors combine to block any vertical flow of air. When a high-pressure system

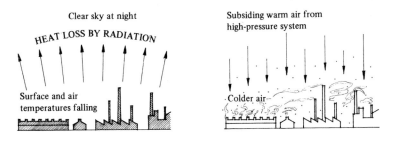

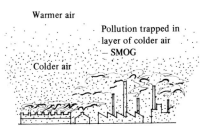

Fig. 3. The combination of temperature inversion and pollution produces smog.

moves over the area it usually brings relatively windless days and the high pressure causes the air to sink. The subsiding air is warmed by compression (just as rising air is cooled by expansion) and the overlying warm air mass traps the layer of cooler air next to the ground. This layer is typically about 2,000 feet thick, but sometimes, as in the case of the London disaster, it can be considerably thinner. Then when poisonous fumes and particulates are emanating from ground-level sources their concentrations can build up very rapidly. Geographical factors also contribute; valleys and low-lying land ringed by hills are especially prone to inversion episodes. Unfortunately, many major cities are thus situated, on rivers or on harbours backed by higher land.

Once an inversion of this kind has occurred it is augmented by the presence of pollutants in the atmosphere. Smoke and particles cut down on the amount of sunlight reaching the ground and the particulates act as nuclei for the formation of fog. Then the sun's energy cannot penetrate as effectively to warm the ground and break up the inversion. As the smog builds up, the condition feeds upon itself and the inversion can persist for several days until another weather system moves in, sweeping the stagnant air mass before it.

Ironically, this kind of dangerous inversion situation occurs during atmospheric conditions that would normally bring fair pleasant weather. High barometric pressure is associated with clear skies, gentle winds, and no precipitation. It brings delightful summer days when – in pasture and on mountainside – the air is clear and calm. Only the lightest of breezes stir the mobile leaves of the aspen or the nodding foxtail grasses of the meadows and the sun beats down out of an almost cloudless sky. The earth, warmed by its rays, pours forth a fragrance of wet loam and sweet clover and new-mown hay. In the fall a slow-moving high-pressure system brings the golden days of Indian summer. And in winter after a storm has passed, the winds subside and the world is all blue and white, frozen and shining like a crystal.

These most beautiful of earth-days are dreaded now in the places where people and industry are concentrated. Men shut themselves behind closed doors and pray for a wind to sweep the skies clean or a rain to wash out the poisons that they have poured into the air above their cities. Nor is it just the urban areas. As populations have grown and industry has moved out into the country, the sources of air pollution have become more widespread. The plumes of poisoned air are often carried on the wind into neighbouring counties and states without a significant dilution of the toxic chemicals. Air pollution episodes now frequently extend over thousands of square miles.

An important part of man-made pollution is caused by the burning of coal and oil. In the combustion of these fuels carbon dioxide and water are formed, but burning is often incomplete and small drops of the un-burned fuel are borne aloft, causing a stream of dark smoke. Tiny particles of carbon and ash are released. Depending on the composition of the fuel, various compounds form in the combustion process. They include some organic solids which are known to be cancer-causing agents and mercury compounds which cause damage to the central nervous system. Thus a wide assortment of chemical substances are present in the cloud of particulates or *aerosols*, which is another name for this heterogeneous collection of small bits of solid matter and drops of fluid suspended in the air.

Certain types of coal and oil contain significant proportions of sulphur; when they are burned sulphur dioxide is one of the most important effluents. The combustion of bituminous coal, for instance, can produce sulphur dioxide at the rate of one ton for every ten tons of coal consumed. Sulphur dioxide is normally present in the atmosphere in very small concentrations. (It is one of the gases emitted in volcanic eruptions.) When this gas combines with water vapour in the presence of sunlight, it forms sulphuric acid, which is a very corrosive and irritating chemical, causing discolouration and deterioration of fabrics and paints, corroding steel, dissolving limestone and marble. Priceless works of sculpture and architecture in many cities have suffered severe damage from this powerful acid.

When people breathe air containing sulphuric acid in aerosol form, their bronchial and lung tissues become inflamed and less able to transmit oxygen. As breathing becomes more difficult, the heart is forced to work harder. It is also known that sulphur dioxide combined with other aerosols is much more efficient in causing biological damage than either of these factors alone. The exact mechanism of the chemical reactions is not completely understood but some catalytic processes are probably involved. This combination of particulates and strong sulphur dioxide fumes caused thousands of deaths in air pollution disasters like the great London smog of 1952. Similar killer smogs occurred in the Meuse Valley in Belgium in 1930 and Donora, Pennsylvania, in 1948. In addition to those killed outright, untold numbers of people have had their lives shortened and made less vigorous by frequent exposure to these man-made additions to the earth's atmosphere.

After the principal villains of the killer smogs had been identified it was not very long before methods for removing both particulates and sulphur dioxide from factory emissions were developed. But the problem is not just a technical one. There is still considerable argument about whether it is 'economically feasible' to install these devices on all power plants and factory smokestacks. A less effective method has been very widely used – the building of tall smokestacks which eject the effluents so far above the earth that they are not as frequently trapped in the shallow inversion layers. But from these stacks the sulphur dioxide and the aerosols are still emptied into the atmosphere; they are just dispersed more widely over the countryside and, entering the prevailing wind patterns, they are carried great distances around the world.

At about the same time that the dangers of coal smoke were being dramatically demonstrated in London, scientists in California were becoming concerned about the unpleasant blanket of fog that frequently settled over the Los Angeles basin. This smog was not exactly like the kind that caused the London pea-soupers. It had a different colour and a different smell. Instead of being dark yellow this smog had a faint brownish tinge and a sharp, acrid odour. A Dutch scientist, A. J. Haagen-Smit, who was doing research at the California Institute of Technology, noticed this strange odour and recognized that it was not characteristic of coal smoke. Haagen-Smit was very sensitive to subtle differences in odours; you might say he had an unusually well-educated nose, because for the past few years he had been studying the chemicals that control odours in plants and had been trying to answer such questions as why onions smell different from pineapples. But the odour that he noticed in the Pasadena atmosphere was not a sweet fragrance like pineapple; it was penetrating and pungent like chlorine.

Haagen-Smit collected samples of the air and made a chemical analysis of them in his laboratory. They proved to contain a number of highly reactive compounds – aldehydes and peroxides. As Haagen-Smit had suspected, these were not usual products of factory smokestacks. Many more years of investigation finally led to the discovery of a different source of pollution and a different chemical species to cause concern. Ozone was found to comprise about 90 per cent of the polluting chemicals present in this type of smog.[6]

Most of the ozone in the earth's atmosphere is contained in a layer high up in the stratosphere where it is formed by the action of powerful rays of sunlight striking oxygen molecules and splitting them into single oxygen atoms which then combine with other oxygen molecules to produce the three-atom molecule – ozone. At this altitude ozone performs an important screening function, filtering out the ultraviolet portion of solar radiation, but when ozone comes into contact with living organisms it is very poisonous, more so than cyanide or strychnine. The protective enzyme mechanisms which guard living organisms against the reactive nature of free oxygen are not effective against its more violent cousin, ozone. By its very nature, evolution can respond significantly only to conditions that the organisms experience regularly over long periods of time, and in the lower atmosphere ozone normally exists only at great dilution because the sun's rays which strike the earth's surface no longer have sufficient energy to split the oxygen molecule and thus start the process that leads to the formation of ozone.

However, the formation process can be brought about at ground level by certain catalytic reactions as well as high-energy phenomena such as lightning and electrical discharge. According to the theory worked out by Haagen-Smit, two common types of chemicals acting together can catalyse the production of ozone near the earth – nitrogen oxides and hydrocarbons. Whenever very high temperature combustion takes place in air, nitrogen is fixed, the principal compound formed being nitric oxide. As we have seen, lightning is the largest natural source of this fixed nitrogen, but human activities also create a considerable amount of nitric oxide – automobile engines, steam boilers for generating electricity, and a large number of industrial processes that take place at temperatures high enough to cause the formation of nitric oxide. In itself this compound is not particularly harmful, but when combined with air and hydrocarbons (also present in automobile exhaust) nitrogen dioxide is produced. This oxide is a much more dangerous compound of nitrogen. It has a bitter smell and a whiskey-brown colour which darkens the smog that contains this pollutant.

The third ingredient which is necessary to turn the chemical brew into ozone is sunlight. Although not strong enough at ground level to convert oxygen directly to ozone, sunlight is sufficiently powerful to split one atom of oxygen from the nitrogen dioxide molecule. Once these single oxygen atoms exist they combine rapidly with regular oxygen molecules to produce ozone. This photochemical (or light-energized) reaction also results in the re-formation of nitric oxide, which is then available to enter the cycle again. Thus a self-sustaining chain reaction is set up.

This theory helps explain many of the observations about ozone pollution. The greatest concentrations occur on hot summer days and they reach a maximum in early afternoon, when the sunlight is the strongest. By this time the air has been filled with the exhaust fumes from automobiles and trucks. The power plants, working at top capacity, have poured forth generous quantities of nitric oxide. Printing plants and dry-cleaning establishments – as well as cars – have added their contribution of the catalysing hydrocarbons. And if a temperature inversion exists, the mixture is concentrated in a shallow layer. The sun beating down upon this layer completes the stewing of the witches' brew. In the evening, when the traffic has subsided and the sun has gone down, ozone levels usually drop back to near normal levels. All of these facts are well explained by the theory.

There are, however, some facts which it does not explain. Ozone concentrations are often higher in the country than they are in centres of population and these country areas may be far from major highways or large industry. Occasionally night-time concentrations do not drop back to normal but remain high. In cities like Los Angeles it was discovered that a reduction in hydrocarbon levels did not reduce ozone levels. Furthermore, the days of highest ozone pollution usually occur in late July or August instead of in June, when the sunlight in the Northern Hemisphere is at its peak intensity.

It is apparent that only a portion of the truth has been discovered about the process of ozone formation near the earth's surface. The chemistry is undoubtedly more complex than the cycle of reactions outlined above. Many scientists are working full-time to unravel these problems because the concentrations of ozone to which people are exposed today exceed the danger levels on many summer days and these concentrations seem to have been increasing throughout the last decade.

Unlike the pea-soup smogs of London, ozone pollution is not immediately apparent to the casual observer. At ordinary temperatures pure ozone is a pale blue gas, but in the concentrations normally observed on earth the colour is not noticeable. Its pungent odour can be detected at very low concentrations (the word ozone comes from the Greek word *ozein*, to smell). But after people have been exposed to it for some time their perception of the smell diminishes. The odour, therefore, does not serve as a reliable warning of rising ozone levels.

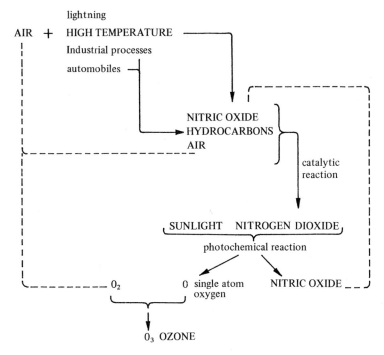

Fig. 4. Ozone pollution – diagram of the cycle of reactions.

Damage to vegetation from exposure to ozone and its close relatives (which together form a group of compounds called *photochemical oxidants*) has been observed for several decades, but the cause was not immediately identified. In the 1940s tobacco growers in Virginia began complaining of mysterious damage to their crops. The undersurfaces of the leaves of the plants developed slightly waterlogged or bruised-looking areas. Then these areas dried out, leaving brown spots of dead organic tissue. Within a few years complaints about similar damage began to pour in from other parts of the country. In California grape leaves were 'stippled' with dark spots; orange and lemon trees lost some of their leaves; and the yield from the orchards and vineyards was reduced. In New Jersey cucumber plants were flecked with brown, and in Wisconsin onion plants were dying back at the tips. Commercial growers of Christmas trees in a 300-

square-mile area in Maryland and West Virginia reported that 750,000 trees were in jeopardy. In the San Bernardino Mountains near Los Angeles magnificent stands of tall ponderosa pines were turning yellow.[7]

At first these symptoms were diagnosed as fungus diseases or insect invasions. Then, after Haagen-Smit's researches on the chemistry of smog in Los Angeles, the true nature of these attacks on vegetation was recognized. Controlled experiments in greenhouses later confirmed the diagnosis. It was definitely established that ozone in concentrations over 50 parts per billion for an eight-hour period causes visible leaf damage to sensitive plants. The colour of the leaf changes from green to yellow and this change occurs in the palisade cells which contain chlorophyll. These facts suggest that the ozone interferes with the process of photosynthesis, and several scientific studies have shown that the rate of photosynthesis is significantly reduced by exposure to ozone. In experiments conducted at the School of Forestry at Yale University the exposure of white pine saplings to consecutive doses of 500 to 800 parts per billion* for three hours reduced their rate of photosynthesis by approximately 80 per cent. Smaller doses also produced measurable reductions in this rate. The scientists conducting these experiments concluded that damage to non-woody vegetation can occur whenever ozone concentrations are greater than 50 parts per billion for one hour or more. It is estimated that this level is exceeded on approximately one third of the summer days in the mid-western and eastern parts of the United States.[8]

Studies with laboratory animals have yielded similar results. When rats were exposed to doses of 100 parts per billion, repeated daily for a year, their lung tissues underwent permanent changes similar to those found in emphysema and fibrosis. Shorter exposures – one hour once a week for fifty-two weeks – resulted in signs of accelerated ageing and weakening of the animal's resistance to bacterial infections of the respiratory system. The biologists reporting these results summed them up in these words: 'Mortality is enhanced . . . by ozone exposure.'

In these experiments with animals the ozone appeared to work primarily upon the respiratory system, and this observation is borne out in the few studies which have been made on human volunteers or humans inadvertently subjected to elevated ozone concentrations. Subjects showed decreased vital capacity in the lungs, accompanied by chest

*A United States billion is equivalent to a British milliard.

pain and cough. Drowsiness, headache, nausea, and inability to concentrate have also been reported after very brief exposures. When the air temperature and humidity are high these effects are magnified; exercise and physical labour bring on the symptoms much more rapidly.[9]

Other more serious types of damage from long-term exposure are suspected. A number of experiments have shown increased rates of mutation, reduced reproductive capacity, and signs of premature ageing in laboratory mice and rabbits subjected to chronic exposure at doses within the range of concentrations which occur frequently in the United States in the summer. Carcinogenic properties are generally believed to be linked to mutagenic effects and studies have shown a high correlation between air pollution and cancer.

As a result of this scientific information, air-quality standards for photochemical oxidants were adopted in the United States, setting maximum permissible levels. Ozone is the principal member of this chemical species. According to the federal standards, 80 parts per billion should not be exceeded more than one hour a year.

However, the discovery of ozone pollution is so recent that adequate monitoring and reporting of these conditions has not yet been established throughout the United States, and in many places where regular information has been collected air-quality standards are often exceeded. In Los Angeles the maximum concentration of 80 parts per billion was exceeded on 237 days during 1974.[10] The city of Chicago began measuring and reporting ozone levels in July 1974, and during the next two months the concentrations were higher than 80 parts per billion on eleven days. In 1975 the standards were exceeded on 28 days. Levels up to 234 parts per billion were recorded. In 1977 standards were exceeded on forty-two days. In 1979 the Environmental Protection Agency relaxed its standard by approximately 50 per cent to 120 parts per billion. This action, taken in response to pressure from the industry, weakens the legislation, which was already inadequate to protect the public. On days when concentrations exceed 70 parts per billion a warning is issued advising people with respiratory or heart ailments to avoid unnecessary exposure to the outside air and any activity involving exercise or physical labour.

Although people with heart and lung problems are certainly the most susceptible to adverse effects from such exposure, there is clear scientific evidence that frequent exposure to such high levels of ozone affects the health of everyone – the young and vigorous as well as the old and sick. This fact is generally acknowledged by air pollution experts, but the official attitude is that the public should not be alarmed. As in the great London smog, people are not properly informed. If they were, they could take protective action, avoiding unnecessary exposure. They could take political action also, demanding better enforcement of pollution standards, better collection of data, and accelerated research into the causes of this new environmental hazard.

Some of the man-made sources have been quite positively identified. Automobiles and trucks appear to be the most important source, since they produce very large quantities of hydrocarbons from incomplete combustion of the fuel and nitric oxide from the burning itself. Fumes from fossil-fuel power plants have been found to cause elevated ozone levels as far as fifty to a hundred miles downwind of the plants. High-voltage electrical equipment such as transmission lines and transformers also cause direct formation of ozone without any catalytic action being involved. All of these sources are increasing in number and size every year. Although methods for reducing the chemical pollution from these sources exist, they have not been consistently applied.[12]

It is conceivable also that changes in natural systems may be affecting ozone levels. For example, alterations in circulation patterns within the earth's atmosphere could bring about a redistribution of ozone, but this possibility cannot be tested until much more information has been collected and analysed.

In the meantime, during many of the brightest summer days when ozone levels are high, people should not engage in active outdoor sports or work in the fields. They should stay indoors, cut down on physical activity, and wait for rain to wash the polluting chemicals out of the air.

When rain comes, the ozone levels drop precipitously. So do the levels of other contaminants. But when rain serves the purpose of scrubbing the atmosphere clean, what happens to the rain? All of the principal pollutants we have been discussing are soluble in rainwater. Sulphur dioxide, nitrogen dioxide, and ozone, when reacting with water vapour in the presence of sunshine, create powerful acids – sulphuric acid, nitric acid, and hydrogen peroxide. Tiny drops of these acids are carried as aerosols on the prevailing winds and are eventually precipitated out far from the original source.

The occurrence of acid rainfall has been observed recently in many places around the world, and concern about this phenomenon is growing. It was first detected in Scandinavia, where measurements have been made regularly since 1956. During this time there has been a 200-fold increase in average acidity of precipitation. Careful monitoring and analysis indicates that most of the pollution problems in Scandinavia come from industrial smokestacks in England and the Ruhr Valley in Germany. Rain in north-western Europe and the entire eastern portion of the United States has also shown increasing acidity over the past fifteen years. Even in such unlikely places as the South Pacific, western Alaska, and the Amazon Basin, acid rain has been collected.[13]

There are almost as many unanswered questions about the phenomenon of acid rain as there are about ozone episodes. How much acidity is caused by sulphur dioxide? How much is caused by nitrogen oxides and other chemicals? In spite of the fact that sulphur content in rain has been somewhat reduced in certain areas such as upper New York State, the acidity of the rain in that region has been increasing.[14]

Does ozone in the atmosphere at ground level increase acidity in rainfall? The locations in which acid rain is found correlate well with areas of high ozone levels.

Other possible contributing factors which are now being studied include the nitrate fertilizers which are spread liberally on agricultural land to encourage rapid growth of crops. The change in the balance of the nitrogen cycle caused by this artificially fixed nitrogen is still undetermined, but some scientists believe that it is contributing to the increasing acidity of rain.

Acid precipitation is very destructive to materials. It eats holes in nylon stockings and causes deep pits in stone and metal structures. The marble Lincoln Memorial in Washington, D.C., for instance, has suffered irreparable damage. One official was quoted as saying that the building seemed to be dissolving like a giant Alka-Seltzer tablet. 'You can almost hear it fizz when it rains.'[15] The spectacular Gothic cathedral in Cologne has been gradually eaten away by acid. Angels have lost their wings and noses and the fine fretwork of marble spires and buttresses is crumbling into dust. Heroic efforts are being undertaken in an attempt to arrest the dissolution of this outstanding example of mediaeval art.

There is a growing body of evidence that acid rain is also damaging to many types of living things. Fish seem to be particularly susceptible. In Scandinavia salmon and trout have disappeared from many of the lakes and restocking has been useless. Thousands of lakes in Norway are now totally devoid of fish. Laboratory experiments indicate that acidity above a certain level kills the fish eggs and fry. The critical level varies for different species. Salmon have a very low tolerance; so do sea trout and brown trout.[16]

Acid water in lakes and rivers causes a shift in the equilibrium of the freshwater ecosystems. Leaching of chemicals such as aluminium, calcium, and magnesium takes place more rapidly. Fungal growth accelerates but bacterial decomposition slows down, increasing the accumulation of organic debris and slowing the recycling of essential nutrients. These changes result in a significant drop in the diversity and abundance of plankton.

The evidence of damage to vegetation from acid precipitation is not as clear-cut. Theoretically the continual leaching of essential plant nutrients from the soil will affect forest growth and other vegetation adversely. To some extent, however, these disadvantages may be offset by the deposition of nutrient elements from the polluted air. Nitrogen compounds, for example, act as fertilizers.

The presumption, however, is that the presence of such acids in the atmosphere, which are washed out regularly year after year into the surface waters of the earth, will eventually affect the balance of the biosphere, favouring some types of life and destroying others. Nature provides water in its most pristine form by evaporation from the lakes, rivers, and oceans around the world. In this purifying process the water molecules are separated from a multitude of adulterating compounds, drawn high into earth's aura, recombined as cloud droplets and snowflakes, and redistributed across the planet's surface, benefiting all living matter. But in passing through the last few hundred feet of air – through man's aura – this purifying process is reversed, and the water that reaches the earth again is a harsh, corrosive element.

Every life form has an aura characteristic of its personality; a breath, a scent, and a flow of energy emanate from it. This luminous cloud may be fragrant and radiant, or sour and dull, depending upon the inner life of the organism. A little perfume or deodorant does not improve an aura for long. Something fundamental must happen before it takes on a more vibrant and pleasing aspect.

Researchers in the Soviet Union and (more recently) in the United States have been using special techniques to photograph the energy field surrounding various plants and animals. They believe they have found evidence that the aura changes colour and shape depending on the health and emotional state of the organism. One curious phenomenon they have reported is 'the phantom leaf'. A photograph of a leaf from which one part has been severed shows faint energy dots still outlining the portion that has been cut off. Only after a lapse of time does the aura assume the leaf's real mutilated shape. Perhaps, these scientists suggest, the aura is an expression of the wholeness – the self – of the organism.[17]

We need not completely accept the scientific validity of these findings to recognize the force of the metaphor. Man's aura is becoming darker and more bitter year by year.

CHAPTER SIX

THE OZONE UMBRELLA

What shelter to grow ripe is ours?
What leisure to grow wise?

           – Matthew Arnold[1]

In the fall of 1974 the newspapers around the United States began carrying banner headlines about a strange new threat to the environment: PERIL TO OZONE LAYER REVEALED; DANGER FROM FREON SPRAY; BAN THE AEROSOL CAN! Most people reading these stories were baffled by the new concepts and long technical words such as 'chlorofluorocarbons' and 'trichlorofluoromethanes'. Even ozone was an unfamiliar term to many people. A few remembered reading recently that ozone 'caused by automobiles' in places like Los Angeles was making people's eyes burn and damaging pine trees in the San Bernardino Mountains. Others, particularly older men and women, remembered vaguely hearing in their childhood that ozone was something good. They had been taken to the country or the seashore in the summertime to breathe the ozone. It was all very confusing. Was ozone good or bad? How could a layer of it floating high in the stratosphere benefit life on the earth's surface? And surely a little squirt from an aerosol can in the privacy of one's own home would be too dilute by the time it travelled fifteen or twenty miles up into the atmosphere to affect even the most fragile of living things back here on earth.

Actually, the significance of ozone in both the lower and the upper atmosphere had come to public attention a number of times over the previous three decades, but none of these earlier news items had generated anything comparable to the excitement that developed about the ozone layer and the aerosol can.

The presence of the ozone layer was discovered when balloon and rocket flights began bringing back information about the atmosphere high above the realm of the clouds and the weather. The temperature inversion at the tropopause (about 6 to 7 miles in mid-latitudes), with the region of rising temperatures extending from there to altitudes of about 32 miles, was a strange phenomenon that needed explanation. In this warm layer, 25 miles or so deep, the atmosphere is extremely rarefied; its density is less than one thousandth of the density of air at sea level. Studies of these tenuous gases revealed the fact that significant amounts of ozone are present. About eight in a million of the widely scattered molecules are ozone. Although only a minor constituent of the stratosphere, ozone is responsible for the rising temperatures at this level. It absorbs the ultra-violet radiation from the sun, converts it into heat and chemical energy, and shields the earth from the full effect of this very powerful radiation.

The chemistry of the gases found in the stratosphere also provided a clue to another mystery about the earth's aura. For many centuries people had been puzzled by the presence of a faint luminous glow throughout the whole sky as though space itself were giving off light. On the clearest of nights when there is no moon, the sky between the stars is very dark but not absolutely black like India ink. All the starlight added together is insufficient to produce the glow which can be seen equally well from vantage points around the whole planet. Its symmetrical distribution suggests something like a spherical halo of radiant space, but it is difficult to see a halo for what it is when one is inside looking out. The shine of it slightly hazes the outline of things beyond the glow, and it is hard to tell whether the haziness comes from the things themselves or from something in between.

By the early twentieth century fine optical instruments had become available for analysing the wavelength of radiation and scientists were able to identify the elements in the atmosphere that were producing the airglow. Oxygen and nitrogen were found to be involved as well as a number of less common elements. We know now that when we look up at the star-studded dome and make out the barely perceptible shining of space between the constellations, we are actually looking at the underside of the ionosphere, a wide, diffuse band of atmosphere where intense solar radiation falling upon the thin gases causes reactions that turn neutral atoms and molecules into *ions* (particles that carry a net electric charge). Many of the reactions also give off light and the emission continues long after the sun has passed into the earth's shadow. In the daytime the light emitted by these air molecules is stronger, but our eyes are so dazzled by other light that we cannot see it. From satellites orbiting in darkness outside the earth's atmosphere, the airglow can be seen and photographed. This luminous halo extends up to 300 or 400 miles; it lies above the ozone layer, because these molecules absorb the high-energy radiation that causes it. The lowest portion of the emitting region is about 55 miles high, producing a very faint but continuous yellow-green light like the cold phosphorescence of fireflies.

In 1956 a group of American scientists working under Murray Zelikoff at the Holloman Air Force Base in New Mexico performed a series of experiments designed to identify the source of this continuous background radiation. In the laboratory they duplicated as closely as possible the conditions thought to exist just above the stratosphere. A large steel sphere was evacuated and filled with a mixture of gases containing oxygen and free nitrogen as well as other trace elements. Then a sunlamp inside the sphere was turned on to expose the mixture to very energetic ultra-violet rays and the results were photographed with a special camera that analysed the radiation produced. From the photographs the scientists could deduce that the artificial sunlight had split many of the oxygen molecules into single atoms. Then another chemical was introduced into the sphere – nitric oxide. Immediately more radiation was produced and the wavelengths of the energy identified the reaction that created it. Nitric oxide combined with single oxygen atoms to make nitrogen dioxide with the release of energy.

Encouraged by this discovery, Zelikoff and his colleagues decided to conduct an experiment in the upper atmosphere to verify their hypothesis that this reaction was one of the sources of the continuous airglow.[3] They had at their disposal a rocket powerful enough to carry a load of chemicals and experimental equipment to an altitude of 60 miles. Two glass vessels containing twenty pounds of nitric oxide were mounted in the head of this rocket and attached to a clockwork mechanism which would release the gas when the rocket reached the top of its trajectory. At 1.45 a.m. on a brilliant starlit March night typical of New Mexico, the rocket was launched. Observers stationed on a hilltop sixty miles away watched with telescopes trained on the portion of the sky where the rocket was expected to release its load of nitric oxide. Two minutes and twenty-five seconds after the launch, a round yellow glow like a misty harvest moon appeared among the stars. It grew in diameter for several minutes until it was four

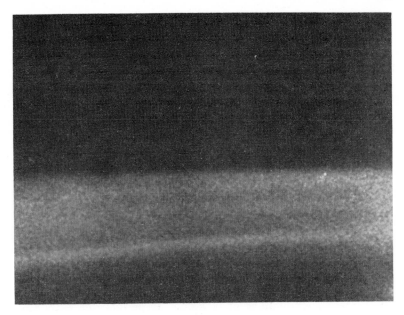

11. The faint luminous halo known as airglow photographed from space.

times the size of a full moon, and as it expanded, the light became fainter, changing colour from gold to silver-grey. Twenty minutes later the light could still be seen, but it was very pale and diffuse.

Zelikoff's experiment sparked considerable speculation about the potential of this man-made moonlight. Perhaps it would be possible to produce the same effect on a grander scale to light whole towns and country areas at night. The maturing of farm crops might be advanced by general night-time illumination. Rescue missions could be carried out at night. Perhaps future space stations could supply all our power needs from this wonderful energy source.[4]

Looking back on these enthusiastic projections with the perspective of twenty years, we recognize that these schemes would have ended in disaster, causing irreparable damage to the band of ozone that shields us from the portion of sunlight that is damaging to life. The illumination of crops at night, far from increasing the yields, would probably have destroyed not only the crops but all vegetation.

Nitrogen oxides floating high above the stratosphere would eventually have drifted down to the ozone layer, where in the presence of ultra-violet light they would act as catalysts, mediating the reactions that destroy ozone, turning it back into normal oxygen. Nitric oxide would be re-formed at the end of the process, making it available again to enter into an endless chain of ozone-destroying reactions. Even a small amount of nitric oxide can produce a large effect because traces of catalysts act like magic wands performing prodigious feats with the littlest of means, just like the ten pounds of nitrogenase that catalyse the total natural fixation of nitrogen by bacteria around the whole world.

It is ironic that nitrogen oxides, which play an important role in the generation of ozone in the lower atmosphere, act as catalysts for the destruction of ozone high above the earth's surface. Versatile villains, these nitrogen oxides, helping to brew up ozone where it is toxic to living things and removing it from the stratosphere where it acts as a beneficial shield for life.

The ozone layer is not a completely uniform stratum, nor does it occur at the same altitude around the globe. Its position follows the same profile as the altitude of the tropopause, lying closest to the earth over the poles and rising to maximum height over the equator. In this region ozone is continuously being made and destroyed by natural processes even without any interference from man. During the day the sun breaks down some of the oxygen molecules to single oxygen atoms, and these, reacting with the oxygen molecules which have not been dissociated, form ozone. However, the sunlight also breaks down ozone, converting some of it back to normal oxygen. And naturally occurring nitrogen oxides enter into the cycle, speeding the breakdown reactions. The amount of ozone that is present at any one time is the balance between the processes that create it and those that destroy it. Like the bank account of a wealthy but extravagant man, the income and outgo are very large but the balance is slim and very sensitive to slight changes in the rate of expenditure.

Since the splitting of the oxygen molecules depends directly upon the intensity of solar radiation, the greatest rate of ozone production occurs over the tropics. However, ozone is also destroyed most rapidly there and wind circulation patterns carry the ozone-enriched upper layers of the atmosphere away from the equator. It turns out that the largest total

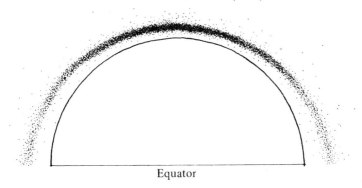

Equator

Fig. 5. The altitude of the ozone layer decreases and the amount of ozone increases with latitude.

ozone amounts are found at high latitudes. On a typical day the amount of ozone over Minnesota, for example, is 30 per cent greater than the amount over Texas, 900 miles farther south. The density and altitude also change with the seasons, the weather, and the amount of solar activity.[5] Nevertheless, at any one place above the earth's surface the long-term averages maintained by natural processes are believed to be reasonably constant. These are the conditions under which life has evolved over millions of years and to which it has become most perfectly adapted.

It may seem strange that the gas which causes health hazards in the lower atmosphere is the same gas that is so beneficial 15 to 30 miles up in the sky. But in biological systems this type of situation is not unusual at all. If we think of the earth and its atmosphere as one giant living organism, then it is easy to understand how the removal of an element from the outer layer of its protective membrane and the injection of that element into deeper layers below the skin could cause serious damage to the organism.

The amount of ozone near the earth is only a few per cent of the amount in the stratosphere, and exchange of molecules between the ozone layer and the air at ground level is thought to be relatively small. Furthermore, ozone is so unstable an element that only a tiny fraction of it could survive the long trip from sea level to the stratosphere, so the ozone layer will not be replenished in any significant degree by the increasing concentrations of ozone that have been detected in recent years near the earth's surface. There is evidence that violent storms may dump some stratospheric ozone into the lower atmosphere, but this effect is transient. The long-term averages of ozone both near ground level and in the stratosphere are regulated by continuous processes that are constantly destroying and creating it in each of these places. The balance may be changed momentarily by a small injection from another source – just as a bank account can be temporarily revivified by the deposit of a birthday cheque; but the bank balance will not be improved for long unless there is a significant change in the regular rate of income and outgo. This is why the presence of catalysts is so important in determining ozone concentrations and why scientists are so concerned about the injection of chemicals like nitrogen oxides into the stratosphere. If the ozone layer is depleted significantly, more ultra-violet radiation would penetrate to the earth's surface and damage many living organisms.

One of the most serious dangers for human health is the increased risk of skin cancer, including malignant melanoma, a rapidly spreading and frightening form of cancer which causes death in about 40 per cent of recorded cases. According to studies conducted in the United States, the prevalence of most types of skin cancer varies geographically; people living in the South have a higher incidence than those living in the northern states, and these observations correlate with the increasing thickness of ozone layers in higher latitudes. In 1974 the National Cancer Institute made a survey of the number of cases of the two most common forms of skin cancer (squamous cell and basal cell carcinoma) in Dallas and Minneapolis. The Dallas figures (4 cases per 1,000 population) were more than twice as large as the number of cases in Minneapolis (1·5 per 1,000), where the ozone layer is known to be about 30 per cent thicker.[6]

Through the process of evolution nature responded to this health threat in the equatorial and tropical latitudes by developing protective skin pigmentation to screen out ultra-violet rays. If the ozone layer undergoes a sudden change, it will be the fair-skinned people who will suffer the greatest damage. Projections vary widely, but it is possible that a 5 per cent reduction in the ozone shield would cause 30,000 additional cases of skin cancer in the United States every year and 500,000 worldwide.[7]

Although the correlation between the amount of ultra-violet irradiation and incidence of malignant melanoma is less well documented, there is good evidence that the disease is linked to exposure to sunlight. The incidence of this very threatening form of cancer is rising rapidly in all countries – with increases of from 3 to 9 per cent a year. In spite of greater vigilance and improved surgical techniques, death rates from the disease have doubled in the last fifteen years. In Canada the disease is increasing at a rate higher than that of any other type of malignancy except lung cancer in males. Some dermatologists believe that the increasing incidence may be at least partly attributed to the fashion for suntans and clothes that give less protection from sunlight.[8]

High-frequency ultra-violet light is also injurious to many other forms of life. Small organisms such as plankton in the sea are very susceptible to increased ultra-violet irradiation. Plankton serves as the base for the aquatic food chain and, therefore, all forms of life in the sea would be affected if the ozone layer were depleted. The tropical seas contain few of these micro-organisms compared with the rich profusion present in the oceans of higher latitudes. In fact, the beautiful clarity of tropical water is a sign of this sterility, which is caused partly by the scarcity of nutrients in the surface layers and partly by the larger component of ultra-violet radiation at these latitudes.[9]

With a less effective ozone shield all life that is exposed to sunlight would undergo faster mutation and genetic defects would be more prevalent. Experiments with plants have shown that greater exposure to the frequencies that are absorbed by ozone leads to stunting of growth, tumours, and abnormalities that are almost always detrimental.

These hazards can be clearly foreseen; other, more subtle changes can only be conjectured. The complexities of the interaction between living organisms and the whole sheltering, nourishing envelope of earth's aura are simply not sufficiently understood today to anticipate all the biological effects of destroying the ozone layer. But scientists are quite well agreed that life in its present form could never have evolved on land without the presence of this protection and, therefore, it is reasonable to infer that the consequences of depleting it would be profound.

Considering the prodigious task that it performs, the amount of ozone in the stratosphere is much less than one might suppose. If it were all brought down to sea level, where it would be subjected to the atmospheric pressures we usually experience, it would occupy a layer only one eighth of an inch thick. So thin a shield protects the whole biosphere – hardly more than the thickness of an umbrella![10]

These relatively few molecules not only intercept the ultra-violet light, they also create the giant inversion layer that inhibits vertical air circulation above the tropopause. Elements introduced into this layer of atmosphere may remain there for years until they are ultimately transported downwards into the lower levels and washed out in rain. A report of the Climatic Impact Committee of the National Academy of Sciences likened the stratosphere to a city whose garbage is collected every few years instead of daily. 'Since man does not control the cleaning operation in the stratosphere, he has no alternative but to live with several years' accumulation.'[11]

Horizontal mixing, on the other hand, takes place quite rapidly within each hemisphere. Substances released in the stratosphere spread in an east–west direction around the world in a week and from the poles to the equator within a few months.

If the ozone layer were depleted, the temperature inversion at the tropopause would be weakened, altering circulation patterns in the upper atmosphere. Less heat would be absorbed by the ozone layer and radiated downwards towards the surface of the planet. The earth's weather would undoubtedly be affected, but no one is sure just what changes would occur or how far-reaching they would be.

We recognize today what a fortunate thing it is that the project of creating night-time illumination for villages and farms by injecting nitrogen oxides high into earth's aura was never brought to fruition. Other scientific dreams, however – even more fantastic ones – were realized, and at the time of their conception nobody anticipated that they might have an effect on the ozone layer.

At dawn on 16 July 1945, only a few miles from the launching place of Murray Zelikoff's experiment, a new weapon was tested in the New Mexico desert at a place called Alamogordo. Brigadier General Thomas Farrell, who was one of about twenty people present, described the event in a report by the War Department to the American people:

The scene inside the shelter was dramatic beyond words . . . We are reaching into the unknown and we did not know what might come of it. It can safely be said that most of those present were praying – and praying harder than they had ever prayed before. If the shot were successful, it was a justification of the several years of intensive effort of tens of thousands of people – statesmen, scientists, engineers, manufacturers, soldiers, and many others in every walk of life.

In that brief instant in the remote New Mexico desert, the tremendous effort of the brains and brawn of all these people came suddenly and startlingly to the fullest fruition. Dr Oppenheimer, on whom had rested a very heavy burden, grew tenser as the last seconds ticked off. He scarcely breathed. He held on to a post to steady himself. For the last few seconds, he stared directly ahead, and then when the announcer shouted 'Now!' and there came this tremendous burst of light followed shortly thereafter by the deep growling roar of the explosion, his face relaxed into an expression of tremendous relief. Several of the observers standing back of the shelter to watch the lighting effects were knocked flat by the blast . . .

The effects could well be called unprecedented, magnificent, beautiful, stupendous and terrifying. No man-made phenomenon of such tremendous power had ever occurred before. The lighting effects beggared description. The whole country was lighted by a searing light with the intensity many times that of the midday sun. It was golden, purple, violet, gray and blue. It lighted every peak, crevasse and ridge of the nearby mountain range with a clarity and beauty that cannot be described but must be seen to be imagined. It was that beauty the great poets dream about but describe most poorly and inadequately. Thirty seconds after, the explosion came first, the air blast pressing hard against the people and things, to be followed almost immediately by the strong, sustained, awesome roar which warned of doomsday and made us feel that we puny things were blasphemous to dare tamper with the forces heretofore reserved to the Almighty. Words are inadequate tools for the job of acquainting those not present with the physical, mental and psychological effects. It had to be witnessed to be realized.[12]

The dangers of the radioactive materials released by an atomic explosion are well recognized, but there is another, more subtle danger. In the fiercely hot crucible of the fireball, free nitrogen combines with oxygen to make nitrogen oxides and the heat of the explosion causes an updraft as powerful as a hurricane directly above the site of detonation. The rapidly rising air current acts like a chimney sucking all the products of combustion high into the atmosphere. The mushroom cloud of the first bombs, such as the one tested at Alamogordo, rose to about five or six miles, but a few years later bombs were developed that were powerful enough to blast the cloud of nitrogen oxides and radioactive dust into the stratosphere.

In the decades following this first blast in the New Mexico desert, numerous atomic tests have been performed and these have added an extra burden of catalytic chemical to the stratosphere. During 1961 and 1962, when the United States and the Soviet Union set off about 300 megatons of nuclear devices, the ozone layer may have been depleted by about 4 per cent. This theoretical estimate, however, has not been verified by actual measurements of ozone depletion following an atomic test.

Even if 'clean' bombs that did not produce lethal radiation were invented, an atomic war in which many of these bombs were exploded might destroy much of the remaining life on earth by reducing the effectiveness of the ozone shield. A study made by the National Academy of Sciences indicated that a nuclear attack involving 10,000 megatons could wipe out as much as 70 per cent of the ozone layer over the Northern Hemisphere and as much as 40 per cent over the Southern Hemisphere.[13]

International agreements have helped to control the number of atomic tests in the atmosphere, although danger to the ozone layer was not the principal reason for these agreements. The first clearly recognized threat to stratospheric ozone was a proposed fleet of high-flying aircraft – the Boeing 2707 supersonic transports. Designed to travel faster than the speed of sound and to operate at altitudes of 10 to 12 miles, these planes would expel large amounts of nitrogen oxides into the stratosphere. The amount produced by such planes depends upon the size and design of the engine. Unfortunately, the efficiency of jet engines is improved when they operate at very high temperatures, thus causing increased generation of nitrogen oxides. The destruction to the ozone layer is also increased when the plane flies at high altitudes.[14]

This hazard to the ozone layer became an important issue before any commercial SSTs were in operation in the United States, and because of the concern expressed by scientists, the appropriations to finance the production of the Boeing 2707s were not approved by Congress. France and Russia have gone ahead with the development of this technology. However, their first SSTs are smaller than the ones proposed in the United States and do not fly quite as high (an average of about 10 miles). Therefore, these planes are expected to have somewhat less impact.

Scientists at Massachusetts Institute of Technology estimated that a fleet of 500 Boeing 2707s if flown at an altitude of 12 miles would cause a 16 per cent reduction of the ozone layer in the mid-latitudes of the Northern Hemisphere, where they were expected to operate. The Southern Hemisphere would suffer an 8 per cent depletion even though it was assumed that no SSTs would fly there.[15] According to the Climatic Impact Committee of the National Academy of Sciences, these figures imply that for each Boeing 2707 put into service a few hundred additional people would probably have contracted skin cancer each year in the United States and several additional deaths would have occurred.[16]

Although more recent studies suggest that ozone depletion caused by nitrogen oxides may not be as great as initially estimated, this issue is still a matter of concern as larger numbers of planes are designed to fly higher in the stratosphere. Future engines operating at higher temperatures would also increase the amount of ozone depletion.

Theoretical possibilities exist for reducing the amount of nitrogen oxides expelled by planes operating in the stratosphere, and laboratory experiments have confirmed the theories. According to the previously cited report of the Climatic Impact Committee of the National Academy of Sciences: 'What needs to be done can probably be accomplished in a decade by a technological society dedicated to this objective. Whether it will be done depends largely on the extent to which the public is prepared to press for it and to pay for it.'[17]

Pressure groups advocating either atomic testing or large fleets of SSTs and military planes argue that the technology they endorse will only add a small amount of catalyst to the stratosphere. But it should be remembered that these technologies are being developed simultaneously and the effects are additive.

In 1974 an even more serious threat to the ozone layer was discovered – the tiny but ubiquitous aerosol spray can. It may seem absurd to compare the impact of the spray can with the SST or the atomic bomb, but it turns out that the worldwide use of the spray can may have a greater long-term effect on the earth's atmosphere than either of those other technologies (except in the case of an atomic war). Each of the aerosol cans is small indeed, but there are billions and billions of them and each one is 'detonated' hundreds of times. These spray cans are a threat to the ozone layer for the same fundamental reason that atomic bombs and S S Ts are. They produce a catalyst that speeds the disintegration of ozone. In this case, however, the catalyst is not nitric oxide but the free atom of chlorine, which is three times more effective in annihilating ozone.

The propellant in approximately half of the aerosol cans is a chemical compound with a long, awkward name – chlorofluorocarbon. Actually there are a number of these compounds belonging to the same family. Du Pont, one of the principal manufacturers, has given them the simpler trade name Freon, but since this name applies only to the Du Pont products, we will use the shortened form of the generic term, *fluorocarbons*, in referring to this group of compounds.

Fluorocarbons have a number of special characteristics which make them very useful. They are extremely inert chemicals which do not react with other substances in the lower atmosphere. They are non-toxic, non-irritating, and they were thought to be remarkably safe to use as propellants for sprays or as the exchange fluid in air-conditioning and refrigeration systems. Although very large amounts of the gases are regularly expelled into the atmosphere from these devices (approximately 1·7 billion pounds a year in 1973), they were believed to be free of any side effects.[18]

However, it occurred to some chemists that this inertness might be altered by exposure to very intense ultra-violet light – just the kind of light, in fact, that radiates on the stratosphere above the ozone shield. This hypothesis was checked in the laboratory and found to be correct. When struck by ultra-violet radiation, fluorocarbon breaks down, yielding, among other products, two or three chlorine atoms. Each of these atoms, acting as a catalyst, transforms one atom of oxygen and one molecule of ozone into two molecules of normal oxygen. In the process the chlorine atoms are released and are ready to perform their catalytic function again.

The inertness which was an environmental advantage at earth level becomes a great disadvantage when fluorocarbon starts its wandering trip up through the atmosphere. Borne on the erratic updrafts that normally mix the air throughout the troposphere, a molecule may take several weeks to reach the tropopause, and many molecules are so unstable or reactive that long before they reach this level they have changed their identity. But fluorocarbon molecules make the trip intact and arrive eventually at the upper edge of the troposphere.

Above this point the temperature inversion stops the general upward movement. The fluorocarbon molecule, however, is in no hurry. Its stable structure gives it a long life expectancy and makes it a patient wanderer. Eventually the circulation patterns of the stratosphere carry it up to heights of approximately 20 miles, where it meets radiation of sufficient strength to decompose it and liberate the free chlorine atoms. Estimates of the average time needed for the fluorocarbon molecule to diffuse from ground level to the ozone layer range from 20 to 50 years.

The same atmospheric conditions that delay the diffusion of fluorocarbon into the stratosphere cause a lengthy residence time for the chlorine atom in the ozone layer. It remains there at least two years, perhaps as long as ten.[19]

Until very recently this scenario was based entirely on theoretical calculations, but now actual data on the chemical composition of the stratosphere are being collected from balloon flights. By an interesting twist of fate, many of these flights designed to diagnose the health of the ozone layer have been launched from the Holloman Air Force Base – the same place where Murray Zelikoff sent off his rocket carrying nitric oxide above the stratosphere; it is within sight of Alamogordo. The information collected so far confirms the fact that fluorocarbon molecules are present at altitudes higher than 12 miles, the quantities being in close agreement with the predictions of scientists.[20] Furthermore, considerable amounts of hydrogen cloride (a simple compound of hydrogen and chlorine) have been discovered in the stratosphere. Some hydrogen chloride is normally present in the atmosphere, derived from salt sea spray, and the scientists found that the concentration of hydrogen chloride decreased steadily up to heights of about 8 or 9 miles, as would be expected with a chemical of oceanic origin. But then above this altitude the concentration began to increase, reaching a maximum at a height of about 15 to 17 miles. The abundance and distribution of this chemical in the stratosphere correlate well with estimates based on its formation from the breakdown of fluorocarbons in the ozone layer.[21]

Because of the protracted time spans involved in the transport of fluorocarbons to the stratosphere and their long residence there, the full effect will be cumulative. Chemists estimate that even if the release of fluorocarbons were halted as early as 1978 the ozone reduction would continue to increase for at least a decade. After seventy-five years only one half of the maximum loss would have been recovered.[22]

In 1976, chemists employed by the companies that make fluorocarbons raised an interesting question. Do free chlorine atoms combine with nitrogen oxides to make chlorine nitrate, thus inactivating simultaneously two of the principal catalysts for the destruction of ozone?[23] Perhaps, they say, fluorocarbons are *good* for the ozone layer because they remove the nitrogen oxides! This claim has not been substantiated by other investigators, however; they say that the mitigating effects of this reaction might make fluorocarbons 50 per cent as destructive as originally computed and the effect may be much less significant. Furthermore, the hypothetical importance of chlorine nitrate formation is based entirely on theoretical models and has not been verified by any actual measurements in the atmosphere.

The argument still goes on; the figures change daily. Studies completed in 1977 indicate that fluorocarbons may deplete the ozone layer even more effectively than estimates in 1976 had indicated.

There are several reasonable alternatives to the present type of spray-can propellant. Many different kinds of fluorocarbons are known and some of them, being less stable, would decompose in the lower atmosphere. One type of fluorocarbon which is suitable for use in refrigeration may be twenty times less destructive of ozone in the stratosphere than the type most commonly used. Although there is some uncertainty in these predictions, the possibility of substituting a less hazardous compound for the various applications is certainly a promising one.[24] Free gaseous nitrogen or carbon dioxide, which are both normal components of the atmosphere, can replace fluorocarbons in spray cans if the consumer is willing to exert a little more pressure with his finger. Manual atomizers and improved pumping devices are also practical alternatives.

In refrigerators and air-conditioning units some improvement could be gained through better engineering. It is believed that the fluorocarbons used in refrigeration in the United States usually get into the atmosphere within two years. This short lifetime as a refrigerant could be very much extended by better design and more durable construction. Automobile cooling systems and cheap air-conditioning units are thought to contribute a large share of this fluorocarbon needlessly dispelled into the earth's atmosphere.[25]

However, as late as 1975, the manufacturers of Freon still maintained that the concern about its effect on the ozone layer was based upon an unproven theory. 'We believe that in the absence of proof of immediate harm,' a quarterly report to their stockholders states, 'Du Pont should oppose any precipitous action that adversely affects the consumer and our business, and in turn damages the nation's economy.'[26]

Scientists questioning the profligate use of fluorocarbons, on the other hand, point out that the very nature of the chemical processes involved makes it impossible to demonstrate immediate harm. The fluorocarbon molecules take 20 to 50 years to reach the ozone layer and the cases of skin cancer which they cause might not show up for another 15 to 40 years after the depletion of the layer occurred.[27] If we must wait until proof of damage is obtained, it will be much too late to prevent serious consequences.

In order to referee this involved argument, the National Academy of Sciences appointed a committee of experts to study all the evidence and make a recommendation for action. Their report, published in September 1976, concluded that a substantial hazard to the ozone layer would develop even if the release of fluorocarbons to the atmosphere were held constant at the same rate as in 1973. In this case, a reduction of 6 to 7·5 per cent would probably occur, resulting in a 12 to 15 per cent increase in ultra-violet radiation reaching the earth. They also noted that although the release of fluorocarbons around the world has levelled off at approximately the 1973 level, during the previous decade it was increasing at 10 per cent a year. This represents a doubling time of seven and a half years. The rate of increase was higher abroad than it was in the United States. Furthermore, the United States manufactures less than half of the fluorocarbons produced worldwide.[28]

The National Academy of Sciences committee recommended that the use of fluorocarbons be controlled in the United States by 1978 unless continuing studies significantly decreased their estimates before that time. Later studies indicated that the committee's estimates were probably conservative and in March 1978 three federal agencies issued regulations bringing about a phased end to the non-essential uses of fluorocarbons in aerosol dispensers by April 1979.[29] This action sets an important precedent which hopefully will be followed by other nations. It demonstrates that it is possible to deal intelligently with our increasing ability to alter the environment.

One of the great advantages man has over other species is to be able to foresee dangers and plan around them. The growing need for this kind of planning has been dramatically brought to public attention by the threat to the ozone layer. Let us hope it will speed the day when the consequences of all new developments will be considered before they are commercially exploited and before a large commitment of time and material has been invested in them.

The threat to the ozone layer is also the clearest example in recent years of an environmental problem that can be solved only by international cooperation. If this protective screen were depleted anywhere north of the equator – over Europe or Russia or the United States – within a few weeks or months the whole Northern Hemisphere would experience increased exposure. No nation, however strong or clever, can alone provide protection for the stratosphere over its land, so the pursuit of national or economic advantage at the expense of the ozone layer cannot be anything but self-defeating. The rich and powerful would be affected just as much as the poor and weak – perhaps even more so, because the white-skinned Caucasian peoples of the mid-latitudes are particularly vulnerable. Nations with a heavily pigmented population and countries in high latitudes would suffer the least.

Science and technology have made it possible for men to disturb the most sensitive parts of earth's aura. They have enabled men to fly planes above the anvil tops of the thunderheads, to lace the upper atmosphere with the white ribbons of jet trails, and to loft rockets into a region of space so charged with energy that it shines with its own fluorescent light. This strange new frontier is almost incredibly susceptible to damage. Pound for pound, contaminants distributed there can do much more harm than they would near the earth because they are reacting with a more dilute form of matter. As in a quiet backwater, pollution collects in the stratosphere and no rain washes it away. Mankind, used to the more forgiving ways of the denser, lower atmosphere, is surprised – perhaps even a little annoyed – to discover the fragility of this layer of earth's aura, which has now become part of the world in which he moves and breathes. The opening up of this new frontier has created an evolutionary challenge demanding a new understanding of planet earth as a whole – a single finely integrated system to be guarded and cared for. Its atmosphere, its oceans, its film of living matter work together like cells in a living organism to produce the beautiful blue planet where flowers brighten the brown crust of the continents, and cloud shadows make moving patterns of light and darkness across the seas, and birds fill the skies with song. The preservation of this delicately coordinated system in the face of increasing human interference and deeper invasion into realms that once belonged only to high-flying cirrus and sunbeams, poses a greater challenge to man's intelligence than he has met since he descended from the treetops and began to fashion his first crude tools.

If human activities destroy the ozone layer, then men will have to retreat to more and more restricted habitats. They will have to take defensive action to protect themselves, devising artificial environments so that they will not be exposed to sunlight. Like the primitive forms of life that inhabited the earth in the Precambrian eras when there was no oxygen in the atmosphere and no ozone to filter out ultra-violet light, they may have to retreat underground, or take up a purely nocturnal existence, or go back to the sea, where thirty feet of water could serve to screen out the lethal rays of the sun.

CHAPTER SEVEN

THE WIND'S WAY

The wind flapped loose, the wind was still,
Shaken out dead from tree and hill.
I had walked on at the wind's will
I sat now, for the wind was still.

                   – Dante Gabriel Rossetti[1]

In the high latitudes of the Southern Hemisphere where the wind blows uninterrupted, a long fetch over thousands of miles of open sea, it piles up towering rollers thirty to fifty feet high, but farther north near the girdle of the earth the winds may be just a whisper – too light to ruffle the glassy surface of the sea. And although many a sailing ship was lost in the foaming, ice-choked seas around Cape Horn and in the violent winter gales that swept the Atlantic, it was perhaps the absence of wind which was most feared. The nightmare of the Ancient Mariner was never far from the thoughts or indeed the experience of the men who depended upon the wind to move their ships along the trade routes of the world.

Alan Villiers, an Australian, was just sixteen in 1919 when he signed on as an ordinary seaman in a lime-juice four-masted bark bound from Australia around the Cape of Good Hope to France. As the ship entered the equatorial zone it encountered the doldrums, where it languished for weeks.

> Day after day the sea lay stagnant all around us and the sails slatted and banged, useless and fretful, against the great steel masts, while the mariners hauled the ponderous yards about to every catspaw that came whispering deceitfully across the oily sea. None grew into usable wind. Some days we moved not at all; on others, backward. Our captain had misjudged the best place to cross the zone of calms which guards the Atlantic sea to ships coming from the south. There were better places to cross, where the northeast trade winds of the Northern Hemisphere could be trusted to dip nearer to the southeast of the Southern, and a

few days catspaw catching would suffice to get a ship across. I did not know that then, but the captain should have known it. I thought three weeks of doldrums was the usual ration with which all sailing ships must contend, and when later we came also into the Sargasso Sea and languished there, I thought that normal too.

We were two months from the Equator to the Bay of Biscay – a longish passage. Day after day in the burning doldrums the sun beat down and the wind, when it came at all, played useless tricks like coming at the ship suddenly from ahead to blow her backward and then, by the time we had all the yards hauled round and the sails properly trimmed, dying at once, so suddenly and thoroughly that it was hard to believe it had ever blown. All the while the Equatorial Current bore us westward, away from our destination, and I began to understand all too well the fears of the ancient mariners who, drifting into these windless zones and observing the current's stealthy, silent thwarting of their every effort to gain ground, held that their ships never could return to the ports whence they had sailed, never could progress again but in one way and that way hopeless.

. . . at last a breeze came which developed into the north-east trades. We were still not finished with windless zones, and after a few days or a week of tolerable sailing, the grass-covered bottom of that steel four-master again languished in almost continual calm. Unlike the dolorous doldrums on the Line, now it was at least dry. Day after day, that 3,000-ton four-master stood upright and idle, mirrored in the blue and slothful water where the lines of golden weed spread out as far as the eye could see, to the right and to the left, in great straight lengths of unblemished perfection which spoke of windless days lasting for weeks, for the slightest ripple on the surface would set the weed to dancing and break the lines.

. . . our captain took to his bunk sick after a week of it, and after a while the sea was so still that scarcely a block creaked in all the maze of our tremendous rigging and the sails hung lifeless, with not even a gentle heave of the ship to send them slatting against the masts. There was not the slightest movement in the oppressive air, and the surface of the sea lay panting quietly, like a great heat-tormented beast. Day after day, we drifted in calm, while the ship stood upright mirrored in her own stagnant image, and cans we threw overboard glinted in the blue depths, down and down and down, a hundred fathoms down, it seemed, where the light still caught them. And our garbage littered our unseen wake, until the sharks came, and the sailors were afraid to swim, and the pitch bubbled in the seams of our wooden deck. Our passage became a drifting match.

We were more than four months at sea before we sighted the island of Flores in the Azores. There we were becalmed again and were another month before we arrived at last at our destination.[2]

Villiers's ship had crossed the two most windless zones on earth – the doldrums and the horse latitudes. Along the equator there is a narrow band where surface winds are characteristically light. In these low latitudes the air, heated by solar energy, rises strongly, causing an area of low atmospheric pressure. Like water flowing downhill, air moves from high-pressure to low-pressure regions. So when hot air rises over the equator, the near-by surface air flows in towards the equator from both north and south to equalize the pressure. But equalization is never achieved because the sun warms the air day after day, causing the continual updraft, and the surface air continues to rush into the resulting low. These steady currents of air towards the equator are the trade winds, which blow very consistently at about 11 to 14 miles an hour. In the region where the trades meet and the flow turns upwards there is a quiet zone known as the doldrums.

Since it is caused by the sun's heat, the doldrums changes its position slightly with the seasons, moving as much as ten or fifteen degrees. Its location is also influenced by the presence of land masses, because land heats up more rapidly than water in the daytime and loses its heat more rapidly at night. Since the Northern Hemisphere contains most of the land mass of the world, the position of the doldrums is displaced northwards. It lies between the equator and 15 degrees north with fluctuations caused by the seasons and differences in local topography. These variations in the position of the doldrums were of great importance to sailing ships and, as Villiers said, the best captains made it their business to know the most favourable places to cross, places where the north-east trades dip nearest to the south-east trades of the Southern Hemisphere.

The equatorial zone receives the maximum amount of solar energy because the incoming rays here are more nearly perpendicular to the earth's surface. At mid-latitudes the rays strike the ground at a glancing angle; at the poles they are almost parallel and have little effect. So the spherical shape of the planet results in an uneven distribution of solar energy and this energy difference is the principal moving force of the wind and weather systems.

The long, warm days of the tropics are punctuated by afternoon thunderstorms and by squalls which occur most frequently at the time of the equinoxes, when the sun shines directly down on the equator. The rising column of warm air is loaded with moisture that condenses out into clouds and the condensation warms the air again. This process breeds thunderstorms which typically occur in clusters several hundred miles across, travelling slowly westwards. These storms act like enormous pumps sending pulses of energy high into the atmosphere, carrying warmth from the sun and moisture from the tropical seas to altitudes of eight to ten miles. Here at the tropopause the updraft ends; the air turns and moves towards the poles.

If the earth did not rotate on its axis, the direction of flow would be directly north and south, but the turning of the earth imparts a curved motion to these winds. To understand why this happens, imagine that a giant standing with his back to the equator throws a ball towards the North Pole. An observer in a satellite hovering in space above the earth would see the ball moving in a straight line. But to an observer on the earth the path of the ball would be curved because it is moving east more rapidly than the ground beneath it. All objects on the equator, including the giant and the ball (before it was thrown), are rotating eastwards at a faster speed than objects farther north or south. In Washington, D.C., for example, the easterly component of motion is less than at the equator and in Montreal it is smaller still. But the ball retains the easterly component of motion which it possessed when it left the giant's hand, and by the time it reaches Washington, D.C., it is moving east faster than the terrain below it. Therefore, in relation to the ground, it curves to the right of its path as it travels north. If the ball is thrown from the equator towards the South Pole, it will also curve towards the east but this time to the left of its path. In the Northern Hemisphere all moving objects, including the winds, are deflected to the right of their trajectory by the earth's motion and in the Southern Hemisphere to the left. (This deflection is known as the Coriolis effect.)

Cooled even further by their passage towards the poles, the tropical air masses begin slowly to sink again. In the meantime, the motion of the earth has deflected them, so that by the time they reach the 30th or 35th degree of latitude the air in both these bands is travelling almost due east. As it sinks it causes high-pressure areas, forcing the air in the lower atmosphere to move out. Nudged by this high pressure and 'attracted' by low pressure near the girdle of the earth, surface winds flow towards the equator, and as they move they are deflected to the right of their path in the Northern Hemisphere and to the left in the Southern. Thus the north-east and south-east trade winds are created, completing the cycle of the circulation of tropical air.

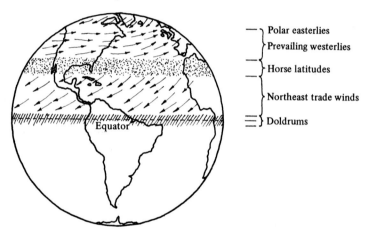

Polar easterlies
Prevailing westerlies

Horse latitudes

Northeast trade winds

Doldrums

Fig. 6. Typical wind flow patterns in the Northern Hemisphere in winter.

The latitudes around 25 to 35 degrees, where the downward air movement is dominant, are characterized by very light winds, and sailing ships like the one that Alan Villiers described were often becalmed there for weeks. Ships that plied the routes between the Old World and the New frequently carried a cargo of horses, and when becalmed so long, they ran short of food and water. Many horses died of thirst and starvation and were thrown overboard. Hence the name 'horse latitudes'.

These regions with their subsiding air masses have almost continual high-pressure conditions. They are fair-weather zones with clear skies and exceedingly low humidity because most of the water is condensed out when the air rises to very high altitudes in the equatorial zones. As it sinks again over the horse latitudes the air is warmed by compression, resulting in a further decrease of relative humidity. The greatest deserts of the world fall into these bands: the Sahara, the Arabian, the Kalahari, the south-western desert of the United States, and the Great Victoria Desert in Australia.

The Sargasso Sea also lies partially within the horse latitudes. It is a vast stagnant region of the Atlantic Ocean, extending south of Bermuda to the Bahamas and east more than halfway across the Atlantic Ocean – a giant backwater, formed by the prevailing ocean currents that encircle it and that feed into it millions of tons of sargassum weed. This brown algae lives attached to rocks and reefs along the shores of the West Indies and Florida. But when it is torn loose by storms, it drifts into the Sargasso Sea, where it floats loose, thriving and multiplying in the brackish sea. This is a place shunned by the winds, where storms and rough seas seldom disturb its stagnant surface. No inflow of fresh water feeds the Sargasso Sea, and it receives a large flux of solar energy causing rapid evaporation and making its saline content unusually high. A profusion of aquatic life inhabits these lush sea pastures. Sea slugs with soft shapeless brown bodies live in the weeds; flying fish make their nests there for the eggs which bear a remarkable resemblance to the round squishy sargassum pods. And eels breed here, returning year after year to lay their eggs in the warm salty waters. One can imagine that just such a scene as this inspired Coleridge's description:

> The very deep did rot: O Christ!
> That ever this should be!
> Yea, slimy things did crawl with legs
> Upon the slimy sea.[3]

Columbus first reported the existence of the Sargasso Sea. His fleet encountered it on his pioneer voyage in 1492, and his sailors were frightened by the eerie expanses of yellowish growth which stretched from horizon to horizon and which seemed to become thicker as they penetrated farther west. The fleet changed course many times searching for a way out of the morass, but the equatorial current and a persistent light easterly breeze carried them relentlessly day after day deeper into the Sargasso Sea. The conviction grew that there would never be a wind to carry them back to Spain. Soon large flights of birds were sighted heading south. 'All night we could hear birds passing,' Columbus wrote.[4] Surely these birds were heading somewhere; there must be land ahead.

It was early October and Columbus was witnessing the annual migration of birds from eastern North America to the West Indies. Some species follow a route which takes them several hundred miles out to sea, even as far as Bermuda. This path enables them to ride the very steady and reliable north-east trade winds south to their winter home in the islands.

Prevailing winds distribute life as well as heat energy in characteristic patterns of scarcity and abundance around the planet's surface. Birds and insects that can soar on the wind move easily from continent to continent and island to island downwind. It is only very rarely that they move in the opposite direction, against the wind. In the north temperate latitudes the flight across the Atlantic from Europe to the United States involves going west against the prevailing winds. Only five species of European land birds have been found in North America, while in Great Britain alone fourteen land birds endemic to America have been seen. On the island of South Georgia in the Antarctic there are two species, a pipit and a teal, whose nearest relatives live 1,000 miles due west to windward on the tip of Tierra del Fuego.[5]

Some of the more earthbound forms of life have invented ways of hitching a ride on the wind – the thistledown, the milkweed seed, and the winged seeds of the maple tree. Spiders and caterpillars ride the winds on their own diaphanous sails. It is a fascinating sight to watch a gossamer spider on an October afternoon, suggests Guy Murchie:

> [Watch him] standing on a tree limb or a gatepost preparing for take-off. With his tiny head to windward he spins out into the sky a fine silken thread from the spinneret at the end of his body. It's a 'run-up' of a special kind. When this spider spinnaker is long enough to hold some real wind and he feels the halyard tugging hard, the spider lets go and sails away on the breeze.
>
> Where to? That is a matter for the wind to decide. However, the gossamer spider has his slight measure of control. If the wind rises, he can reef in his thread to 'shorten sail'. If the wind drops, he can pay it out again like a yachtsman in a tight race. His own body is his craft

and crew. He may sail thus on the wind for hundreds of miles, even across continents or if the wind will it so, rising as high as 60,000 feet and remaining aloft for weeks.[6]

An imaginative scientist calculated the length of cord that would be required to enable a man to float on the wind like gossamer spider and discovered that it would require a strand 124 miles long (fifteen to twenty times the depth of the troposphere). But a hundred strands 1·24 miles long could be arranged on spindles, he suggested; then the pilot could reel in and control his own sails.[7]

On a sunny day in late August or September the air that rises from a meadow may take on almost an iridescent sheen, it is so interlaced with the shiny filaments of webs spun by spiderlings and caterpillar-lings embarking for parts unknown. Aviators have reported seeing great numbers of silky spider webs miles above the earth's surface, and spiders are among the first forms of life to appear on islands newly born out of the sea. Darwin found spiders on St Paul's Rocks, where not even a lichen grows. When Krakatoa erupted in 1883, most of the island itself was destroyed and the small portion that remained was buried deep in hot lava and ash. Scientists studied these remains for signs of surviving life but found none until nine months later the naturalist Cotteau reported that he found one microscopic spider. 'This strange pioneer of renovation was busy spinning his web.' Practically no other forms of life were found on Krakatoa for a quarter of a century; then a number of species began to colonize the island: birds, lizards, snakes, various molluscs, insects, and earthworms. Dutch scientists estimated that 90 per cent of the new inhabitants were forms of life that could easily have arrived by air.[8]

The islands of Polynesia are also believed to have been colonized by air and sea. Spores and seeds and winged forms of life rode in on the trade winds; lizards and tortoises and crabs floated in on natural rafts of tree branches and tangled masses of vegetation. As Thor Heyerdahl proved in his voyage on *Kon Tiki*, the winds and currents bearing westward from South America could very well have brought various species of life to these islands.

Winds also helped to set the patterns of migration of people. Prevailing winds influenced the location of trade routes and the flow of commerce. Favourable weather conditions played an important part in the distribution of populations. It is an interesting fact that some of the most important early centres of civilization sprang up in the high-pressure band of the horse latitudes, where sunshine is abundant and the climate is relatively benign. Dryness – the greatest threat to life in these latitudes – was mitigated in certain especially favoured places by the presence of major rivers – the Nile, the Tigris and Euphrates. In these latitudes, however, the environment is very fragile. Land stripped of its original forest cover remains bare for centuries because the dry conditions make it very difficult for new growth to take hold. According to ancient writers, the lands along the Mediterranean coast near Carthage were once heavily forested. One could ride from Tripoli to the Atlantic in the deep shade of trees. The wild beasts which amused the populace in the Roman arenas came from these Tunisian forests. In biblical times Palestine was a fertile land, too. The hills were clothed with the silvery grey-green of olive groves and a dark mantle of cedar and cypress. As trees were cut down these slopes were gradually stripped bare. Erosion became an increasing problem, destroying the fertility of the land, silting the riverbeds and harbours. [9]

The areas of persistent high pressure are also especially likely to suffer from man-made pollution. Long-term temperature inversion conditions occur, holding smoke, fumes, and particulates close to the ground. Los Angeles (at 34 degrees north latitude) is an outstanding modern example of a city built in a region naturally blessed with a wonderful climate but which has been marred by pollution aggravated by the persistent temperature inversion.

As the populations of ancient civilizations increased and food supplies became more critical, the advantage began to shift to the inhabitants of the temperate zones, where rainfall was more plentiful.

It is, in a way, ironic that the bands of latitudes between 35 and 60 degrees are known as the temperate zones, because these are the regions of the most variable and unpredictable weather conditions with great extremes of both wind and temperature. Within a few hours, or even within a few minutes, the thermometer can plummet as much as 50 degrees. In North Dakota, on the plains of western Canada and the steppes of Russia, the temperatures normally swing between 110°F (43°C) in summer and −50°F (−46°C) in winter with winds up to 60 miles an hour. The great variability of the climate in these latitudes is caused by the alternating domination by polar and sub-tropical air masses.

In the polar regions easterly winds usually prevail at low altitudes. The air which is chilled over the frozen top and bottom of the planet sinks, causing high pressure and forcing surface winds to flow down and outwards towards the mid-latitudes. But at the same time the sinking air masses of the horse latitudes are also causing surface winds to flow away from this high-pressure zone. The flow towards the equator, as we have seen, creates the trade winds, and a flow in the opposite direction carries warm sub-tropical air towards the mid-latitudes. Here it encounters the advancing polar air mass, and a broad band of turbulence is produced in each hemisphere, with warm and cold fronts, moving on prevailing westerly wind currents around the globe. In the upper troposphere this turbulence takes the form of waves travelling from west to east and, occasionally, remaining almost stationary for long periods. In the lower atmosphere the waves and troughs form large swirls and depressions. These are the typical high- and low-pressure systems that troop in an endless procession around these latitudes, producing the wild winter blizzards, the humid heat of summer, the drenching rains, and the destructive ice storms of the 'temperate' zones.

Controlled by the global wind patterns and the rotation of the earth, the low-pressure whirls rotate counter-clockwise in the Northern Hemisphere and are known as cyclones (this is a general term not to be confused with hurricanes that are called tropical cyclones). In the Southern Hemisphere they rotate clockwise. High-pressure areas (anticyclones) are just the opposite, having winds that travel clockwise in the Northern Hemisphere and counter-clockwise in the Southern.

Those of us who have spent most of our lives in the temperate zones, exposed to the whims of high- and low-pressure systems moving from west to east across the continent, are surprised to discover that this is not the usual wind pattern on earth. Across the whole face of the globe easterly winds are the most common.

Although westerlies predominate in the mid-latitudes, the great wind swirls around highs and lows cause winds to blow from every point of the compass. An east wind usually bodes bad weather because it is caused by a cyclone approaching from the south-west.

The position of this turbulent river of air shifts with the seasons. During winter in the Northern Hemisphere the cold masses push farther south and most of the United States, Europe, Russia, China, and Japan are alternately frozen, thawed, drenched, and buffeted in this burbling, swirling stream of air. In the summer the warmer air flow dominates, pushing the meeting point of the two air masses farther north, so that much of the temperate zones enjoy milder and more stable weather.

The mid-latitudes south of the equator are almost devoid of land masses. New Zealand, the tip of South America and Australia, and a few widely scattered islands are the only lands to experience the full force of the southern current of cyclones and anticyclones. Unmitigated by any contact with the slowing effect of mountain ranges or forests, the gales sweep over endless miles of ocean and produce some of the most violent weather conditions regularly encountered on earth. The whalers that used to ply these seas christened the southern latitudes 'the roaring forties, the howling fifties, and the screeching sixties'. The passage around Cape Horn was particularly feared, not only because of the high seas and the extreme cold but because the direction of the wind was unpredictable. It was not unusual to have to make three or four attempts before successfully rounding the Horn. Richard H. Dana, Jr., in *Two Years Before the Mast* wrote what is probably the best authentic description of a passage through these southern latitudes.

On the brig *Alert*, the crew had had a fairly pleasant run down south from California and had been blessed with mostly favourable winds from the west until they reached the 34th latitude south on 19 June.

> There began now to be decided change in the appearance of things . . . the skies looking cold and angry; and, at times, a long, heavy, ugly sea, setting in from the southward, told us what we were coming to. Still, however, we had a fine, strong breeze, and kept on our way under as much sail as our ship would bear. Toward the middle of the week, the wind hauled to the southward, which brought us upon a taut bowline, made the ship meet, nearly head-on, the heavy swell which rolled from that quarter; and there was something not at all encouraging in the manner in which she met it . . . I stood on the forecastle, looking at the seas, which were rolling high as far as the eye could reach, their tops white with foam and the body of them of a deep indigo blue, reflecting the bright rays of the sun. Our ship rose slowly over a few

of the largest of them, until one immense fellow came rolling on, threatening to cover her, and which I was sailor enough to know, by the 'feeling of her' under my feet, she would not rise over. I sprang upon the knight-head, and, seizing hold of the fore-stay, drew myself up upon it. My feet were just off the stanchion when the bow struck fairly into the middle of the sea, and it washed the ship fore and aft, burying her in the water. As soon as she rose out of it, I looked aft, and everything forward of the mainmast, except the long-boat, which was gripped and double-lashed down to the ring-bolts, was swept off clear. The galley, the pigsty, the hen-coop, and a large sheep-pen which had been built upon the fore-hatch, were all gone in the twinkling of an eye – leaving the deck as clean as a chin new reaped, – and not a stick left to show where anything had stood.

Monday, 27 June:

We were sleeping away 'at the rate of knots,' when three knocks on the scuttle and 'All hands, ahoy!' started us from our berths. What could be the matter? It did not appear to be blowing hard, and, looking up through the scuttle, we could see that it was a clear day overhead; yet the watch were taking in sail . . . We did not wait for a second call, but tumbled up the ladder; and there, on the starboard bow, was a bank of mist, covering sea and sky, and driving directly for us. I had seen the same before in my passage round in the Pilgrim, and knew what it meant, and that there was no time to be lost. We had nothing on but thin clothes, yet there was not a moment to spare, and at it we went . . .

Down came the topsail yards, the reef-tackles were manned and hauled out, and we climbed up to windward, and sprang into the weather rigging. The violence of the wind, and the hail and sleet, driving nearly horizontally across the ocean, seemed actually to pin us down to the rigging. It was hard work making head against them. One after another we got out upon the yards . . .

On Monday, 4 July, icebergs were sighted and for three days the *Alert* tacked back and forth to avoid large islands of floating ice.

At the end of the third day the ice was very thick; a complete fog-bank covered the ship. It blew a tremendous gale from the eastward, with sleet and snow, and there was every promise of a dangerous and fatiguing night. At dark, the captain called all hands aft, and told them that not a man was to leave the deck that night; that the ship was in the greatest danger, any cake of ice might knock a hole in her, or she might run on an island and go to pieces. No one could tell whether she would be a ship the next morning . . .

On 9 July, the captain decided he could not safely round the Horn through this dense field of ice. He ordered the ship turned north-north-east to make for the Straits of Magellan.

However . . . the next day, when we must have been near the Cape of Pillars, which is the southwest point of the mouth of the straits, a gale set in from the eastward, with a heavy fog, so that we could not see half the ship's length ahead. This, of course, put an end to the project for the present; for a thick fog and a gale blowing dead ahead are not the most favorable circumstances for the passage of difficult and dangerous straits. This weather, too, seemed likely to last for some time, and we could not think of beating about the mouth of the straits for a week or two, waiting for a favorable opportunity; so we braced up on the larboard tack, put the ship's head due south, and struck her off for Cape Horn again.

In our first attempt to double the Cape, when we came up to the latitude of it, we were nearly seventeen hundred miles to the westward, but, in running for the Straits of Magellan, we stood so far to the eastward that we made our second attempt at a distance of not more than four or five hundred miles; and we had great hopes, by this means, to run clear of the ice; thinking that the easterly gales, which had prevailed for a long time, would have driven it to the westward. With the wind about two points free, the yards braced in a little, and two close-reefed topsails and a reefed foresail on the ship, we made great way toward the southward; and almost every watch, when we came on deck, the air seemed to grow colder, and the sea to run higher. Still we saw no ice, and had great hopes of going clear of it altogether, when, one afternoon, about three o'clock, while we were taking a *siesta* during our watch below, 'All hands!' was called in a loud and fearful voice. 'Tumble up here, men! – tumble up! – don't stop for your clothes – before we're upon it!' We sprang out of our berths and hurried up on deck . . . directly under our larboard quarter, a large ice island, peering out of the mist, and reaching high above our tops; while astern, and on either side of the island, large tracts of field-ice were dimly seen, heaving and rolling in the sea . . . During our watch on deck, which was from twelve to four, the wind came out ahead, with a pelting storm of hail and sleet and we lay hove-to, under a close-reefed fore topsail, the whole watch. During the next watch it fell calm with a drenching rain until daybreak, when the wind came out to the westward, and the weather cleared up, and showed us the whole ocean, in the course which we should have steered, had it not been for the head wind and calm, completely blocked up with ice. Here, then, our progress was stopped, and we wore ship, and once more stood to the northward and eastward; not for the Straits of Magellan, but to make another attempt to double the Cape, still farther to the eastward; for the captain was determined to get round if perseverance could do it, and the third time, he said, never failed.

From a northeast course we gradually hauled to the eastward, and after sailing about two hundred miles, which brought us as near to the western coast of Tierra del Fuego as was safe, and having lost sight of the ice altogether, – for the third time we put the ship's head to the southward, to try the passage of the Cape. The weather continued clear and cold, with a strong gale from the westward . . .

. . . we were soon up with and passed the latitude of the Cape, and, having stood far enough to the southward to give it a wide berth, we began to stand to the eastward, with a good prospect of being round and steering to the northward, on the other side, in a very few days. But ill luck seemed to have lighted upon us. Not four hours had we been standing on in this course before it fell dead calm, and in half an hour it clouded up, a few straggling blasts, with spits of snow and sleet, came from the eastward, and in an hour more we lay hove-to under a close-reefed main topsail, drifting bodily off to leeward before the fiercest storm that we had yet felt, blowing dead ahead, from the eastward. It seemed as though the genius of the place had been roused at finding that we had nearly slipped through his fingers, and had come down upon us with tenfold fury. The sailors said that every blast, as it shook the shrouds, and whistled through the rigging, said to the old ship, 'No, you don't!' – 'No, you don't!' . . .

After about eight days of constant easterly gales, the wind hauled occasionally a little to the southward, and blew hard, which, as we were well to the southward, allowed us to brace in a little, and stand on under all the sail we could carry. These turns lasted but a short while, and sooner or later it set in again from the old quarter; yet at each time we made something, and were gradually edging along to the eastward.[10]

On Friday, 22 July, Staten Island, off the top of Tierra del Fuego, was sighted. The crew of the *Alert* knew that the Cape had been passed. At sundown they had a favourable breeze and the Atlantic Ocean clear before them.

The rigours of a passage around the Horn are more extreme but not different in kind from the hardships that most people in the world endure occasionally from the wind. And perhaps everyone has wished at some time that the winds would be stilled. But imagine what the world would be like without wind. It would be a harsh and barren place. Intense temperature differences would exist between the equator and the poles, between day and night. No rain showers would sweep over the continents. Water, evaporated from the seas during the day, would fall again on the seas at night. Inland areas would be bare rocky wastelands, untouched by soil-building processes. Only small islands might be habitable for life.

In the Latin language there is one word, *anima*, meaning both wind and spirit. And indeed the wind is the invisible force that seems to make the earth come alive. Without wind, all the vivacity of moving light and shadow would be gone from the world – the sparkle of water riffled by vagrant breezes, cloud masses scudding across the sky, and wind waves passing across fields of ripe grain. Many of the familiar background noises would be stilled, too – the hypnotic pounding of surf, the restless stirring of autumn leaves, the sigh of wind in a pine forest – a sound especially loved by the Japanese, who call it the *matsukoza*, the song of the pines.

Men who have gone to the moon speak of the oppressive silence there and the absence of any movement. It is a dead world – not just because there are no living things, but because there is no sound and no wind.

CHAPTER EIGHT

THE SWIFT JET STREAMS

There are more things in heaven and earth, Horatio, than are dreamt of in your philosophy.

– William Shakespeare[1]

It was a cold December day in 1944. The deeply forested mountains of north-western Montana were wrapped in their winter cocoon of snow. Along the valleys of the Flathead River and the Little Bitterroot where the drifts had piled deep, the ranchers were holed up for the winter. Thin columns of smoke, rising high in the still air, marked the presence of their isolated homes. In the small town of Kalispell the streets had not all been cleared yet after last night's storm. The town was quiet except that from far away the monotonous whine of the power saw could be heard. It began, ceased, and then began again. The sharp crack of an axe rang out clear and echoed across the frozen valleys. The loggers were at work on the lower slopes of the Swan Range.

Two of the woodcutters were returning to their work site of the previous day when suddenly they were amazed to see an enormous object festooning the treetops. This strange apparition, which had materialized out of nowhere in the night, looked like a giant bat whose wrinkled wings had become impaled on the sharp branches of the pine forest. On closer inspection it appeared to be a large deflated balloon made out of parchment-like paper. A portion of it had burned, leaving charred edges, and painted along one side was a red emblem – the sign of the rising sun.[2]

The loggers knew what that sign meant and they were frightened. Up to this point the war with Japan had seemed unreal and very far away to these residents of Kalispell. Now suddenly it had plummeted into their back yard.

The two men immediately returned to town and reported what they had seen. Messages were dispatched to the Federal Bureau of Investigation and the Army Air Force, who sent representatives to investigate the mystery. They found that the balloon was indeed Japanese. There was a great deal of information printed on the bag, including the number of hours spent on its manufacture and the date that it had been completed in a Japanese factory just a few weeks earlier – 31 October 1944. Balloon experts were called in and they concluded that the craft could not have flown all the way from Japan. Even with the most favourable winds, it could hardly have travelled more than four or five hundred miles, and Kalispell is 465 miles east of the Pacific coast. The only possible explanation seemed to be that it had been launched from a submarine operating offshore.

The bag had been filled with hydrogen and would have carried about 18,000 cubic feet of the gas – enough to provide sufficient lift to transport several men and equipment for sabotage. Long rope cables dangling from the bag could have supported a gondola to carry the men. But there was no sign of a gondola near the landing site. A box containing radio parts was found near by and a small undetonated incendiary bomb was discovered attached to the side of the balloon.

The residents of Kalispell were asked by the FBI to keep this strange occurrence a secret for national security reasons. It was not until nine days later that the rest of the country heard about it. The FBI announcement on 18 December left many unanswered questions. Had the balloon carried saboteurs? If so, where were they? Speculation ran rife. *Time* magazine suggested that the men could have parachuted to earth from the balloon, which had been designed to destroy itself in the air. *Newsweek* concluded that the saboteurs must have been Germans carefully fitted out with American clothes and credentials. Japanese would have been so conspicuous that they would surely have been identified.

Two weeks later the government admitted that two other balloons had been found, one in Oregon. The location of the other was not disclosed. Actually, this was not the whole story, by any means. It was only the small tip of an iceberg of chilling information almost totally submerged by military secrecy.

The balloon which had landed in Oregon was discovered by a group of children enjoying a Sunday-school picnic on the beach. A large crumpled object lying on the sand dunes attracted their attention. Several of the children poked at it and started to turn over one corner. Suddenly the object burst into flames and exploded in their faces, killing five children and the pastor's wife, who had been in charge of the ill-fated outing.

Even before these events in Oregon and Montana, strange reports had been pouring in to the armed services. Balloons and fireballs had been sighted all along the western part of the United States. A rancher's wife in Colorado had seen a brilliant ball of fire, about the size of a full moon. 'It came out of nowhere,' she said. 'Just blazed out there in one place, stood for a moment without moving, then vanished.' Similar reports described great silvery balloons that suddenly blew themselves to pieces. Some of the balloons were seen releasing objects high in the air. Bomb blasts had been heard in remote and widely separated places like Thermopolis, Wyoming, and Ventura, California. In downtown Los Angeles tattered bits of rice-fabric paper were picked up on the streets. Reports from Mexico, Canada, and Alaska mentioned seeing as many as seventeen balloons in one day and many of them had dropped incendiary bombs. Throughout the western states forest fires suddenly became so numerous that the Army was called in to help fight them.

Another piece of information, which at first appeared to be unrelated, was brought back by pilots returning from bombing missions over Japan. They described a strange new type of 'anti-aircraft device' – large shining spheres that floated high over the cities. But these spheres did not seem to be very effective in intercepting the enemy planes. They rose to heights greater than the altitudes normally used by bombers and many of them appeared to be drifting out to sea.

The military establishment in the United States did not believe that the balloons found in this country could be the same ones that were seen floating over Japan. It seemed beyond any stretch of the imagination that they could have travelled such immense distances in such a short time. The most widely accepted theory was that balloons were being launched from submarines operating in the Pacific. However, the question of their launching site and range might be definitively answered if a balloon could be recovered intact before it had detonated its charge. Military pilots pursued and attempted to shoot down balloons that they sighted. But – like Halloween ghosts – the balloons seemed to be unaffected by bullets. They sailed majestically onwards even after they had been riddled with holes.

Finally, in January 1945, an Air Force pilot manoeuvred a balloon to earth on a mountainside near Alturas, California. Although he was vividly conscious of the fact that at any moment the balloon might explode and turn his plane into a flaming torch, the pilot flew for hours in tight circles around the balloon, turning his plane in such a way that the air streams from his wings and propeller made a downdraft that forced the balloon to descend. Ground crews were standing ready to capture the enormous craft when it came within reach, and they moved it to Moffett Field near Palo Alto, where dirigible hangars provided enough space to house it.

A study of this captured balloon revealed the details of its construction and confirmed the fact that it must have been launched in Japan. Eventually, after the end of the war in August 1945, the whole story of the great Japanese balloon offensive became a matter of public knowledge.[3]

Sometime earlier, probably in 1942, the Japanese had discovered a remarkably swift current of air that blew eastwards at great heights over their islands straight towards the west coast of America. This current may have been encountered by their bomber and fighter pilots heading for Hawaii and Midway. If so, it was a carefully guarded secret because of its military importance. This wonderful natural super-highway in the sky could reduce the flying time from Japan to the United States by six or seven hours and save thousands of gallons of fuel for one plane alone. Better still, it could carry a balloon offensive to the American mainland without using any fuel at all and without endangering the lives of any Japanese pilots.

During the last year of the war the Japanese carried out this imaginative scheme. They constructed special balloons of long-fibre rice paper, filled them with hydrogen, and equipped them with an ingenious device which regulated their altitude so that a balloon launched at a height of 30,000 feet over Japan might stay within the jet stream until it reached the coast of the United States. When the altitude, measured by a barometer, dropped below 29,000 feet an automatic mechanism released one of twenty small sandbags. If the balloon rose above 36,000 feet a valve triggered by the barometer released some of the hydrogen gas so that the balloon would sink again. Since the jet stream was always found within this band of altitudes, the balloon should remain in the stream and would ride its current for 5,000 miles to America. When it reached the coast of the United States it was programmed to start dropping its cargo of incendiary and anti-personnel bombs. As the last bomb was dropped a block of picric acid was designed to explode, destroying all the instruments, and a pouch of magnesium flash-bulb powder flared up, turning the balloon into a ball of fire.

The Japanese had invented a remarkably clever weapon, but they made one mistake. The electronic equipment was powered with a wet-cell battery which froze in the sub-zero temperatures of the jet stream and only about one tenth of the 9,000 balloons released in Japan ever reached the United States at all. Even with this serious defect the balloon offensive was successful enough to give the U.S. military establishment a very bad scare. Some members of the High Command feared that these aircraft would soon be used to carry bacteriological warfare to the United States.

The voluntary censorship of the press and radio in the United States and Canada was largely responsible for defeating the balloon offensive. Only two or three balloon sightings were reported by the media. The Japanese General Staff believed that of the 9,000 launched in Japan these were the only ones that reached America and after six months they discontinued the operation.

In the meantime the jet stream had been discovered by American pilots. On 24 November 1944, the first important high-altitude bombing mission was undertaken, using a hundred and eleven B-29 planes (the Superforts). The plan of this mission, which bore the code name San Antonio, was to bomb industrial regions near Tokyo during daylight hours. These areas were heavily defended, but the very high altitude that could be achieved with the Superforts protected the bombers from anti-aircraft fire. The bombers took off from Saipan in the Marianas, detoured west of the Bonin Islands to avoid a typhoon which was reported in that region, then headed towards Japan. As the planes approached Honshu, flying between 27,000 and 30,000 feet, the westerly wind increased dramatically, impeding their progress. They had been instructed to bomb their target on a west-to-east axis and as they turned east near Tokyo they were suddenly swept forward in a 150-mile-per-hour gale. With ground speeds of about 445 miles per hour, the bombardiers were not able to compensate for wind drift. Only sixteen out of the many hundreds of bombs dropped hit the industrial target area. The rest missed the factories and fell on the city or into the sea.

On subsequent occasions Air Force pilots reported that flying west towards Japan they encountered such winds that their planes seemed to stand still in the air. Islands that should have been passed long ago remained stationary below them as though the whole scene had frozen. Although the engines were racing at full throttle, the planes were not going anywhere. In some cases the pilots had to turn back before their load of fuel was exhausted. On other occasions, in perfectly clear weather conditions and without any warning, planes had been suddenly wrenched and hammered with a series of violent vibrations as though an unseen hand had picked them up and shaken them in mid air. All of these mysterious experiences occurred in a sharply defined band of space, just at the edge of the stratosphere. When the planes descended a few miles, the wild winds diminished rapidly and the turbulence disappeared.

After the war scientists from all over the world followed up these interesting discoveries with a cooperative international effort to reveal the secrets of these remarkable currents of air and to find ways to use them to advantage. A network of instrumented balloon flights and specially equipped aircraft probed the upper troposphere and the lower stratosphere, plotting the position and the speed of wind streams. It was found that there are seven characteristic jet streams (although they do not all occur at the same time of year) and that they follow undulating courses around the earth at altitudes of six to thirty miles.[4]

Jet streams exist where strong temperature differences occur on long fronts. The four principal jet streams are located near the tropopause, a point of abrupt temperature change. Along these meeting places of hot and cold air masses, the air flows from the warm to the cold side, trying to equalize the differences. The dynamics that create the fantastic speeds in the jet streams are quite complex, involving the rotation of the earth as well as the pressure gradients associated with the tropical and polar air masses. These interacting forces produce tunnels of very rapidly moving air. The fastest winds are in the centre of the tubelike space, which is not round but shallowly elliptical in cross section, having a width of several hundred miles and a depth of only two or three miles. Wind speeds at the edge of the tube are about 50 miles an hour, but they increase rapidly towards the core, where velocities as high as 300 miles per hour have been measured. These velocities are several times greater than those of hurricane winds, which can reduce cities to a pile of rubble and strip the leaves from a forest. The force of the jet winds, however, cannot be directly compared with the winds on earth because the air at these altitudes is much thinner. There are fewer molecules in a given volume of space and the energy the wind carries is correspondingly less.

The relative positions of the jet streams and their change with the seasons follow a simple and logical pattern. Over the equator the tropopause occurs at an altitude of about ten miles. Its height decreases with increasing latitude and over the poles it is approximately six miles high. But the tropopause is not a single continuous layer. There are breaks in it which are located at 30 to 35 degrees in the horse latitudes and around 55 to 65 degrees, where the polar air begins. Jet streams occur at or just below these breaks in the tropopause.

The jet that flows in the break between the mid-latitudes and the polar region is known as the *polar-front jet*. This is the stream that the Air Force bombers encountered between Japan and the United States, the current of air that was supposed to carry 9,000 fire balloons across the Pacific. It is the lowest of the jet streams because the level of the tropopause at this point is only six or seven miles high. It is not surprising, therefore, that this was the first jet stream to be discovered. The winds of the polar-front jet travel from west to east, following a sinuous wave pattern around the globe, influenced by temperature differences in the upper air over land and sea. In summer, when tropical air predominates in North America, for example, the polar-front jet moves to higher latitudes, crossing North America in Canada. In winter it moves farther south, traversing the U.S.A. Its movement with the seasons corresponds with the seasonal shift of the band of turbulent low-atmosphere westerlies. In fact, these two phenomena are now believed to be different aspects of one system.

The cyclones and anticyclones which troop across the mid-latitudes from west to east are triggered by changes in the velocity of the jet stream. Pulsations in wind speed within the core of the jet move slowly along the stream's path, migrating eastward at an average speed of twenty-two miles per hour. Increases in speed cause air to flow out of the stream, just as a brook might overflow as it picks up speed going downhill. Then the vertical column of air below that point experiences a decrease in pressure; the barometer drops and a cyclone develops at ground level. Conversely, when the jet stream slows down, air flows back into it, causing higher pressure under that point on the surface of the earth. This high pressure creates an anticyclone. And the whole system of highs and lows moves slowly eastward.

The *sub-tropical jet* is also an integral part of this system. It flows in the break in the tropopause between the mid-latitudes and the tropics, but it occurs only in the winter season, when the most extreme temperature differences exist between the equator and the poles. It moves at a higher altitude than the polar-front jet because in these latitudes the tropopause occurs at an altitude of about ten miles. This rapid river of air also follows a west to east direction in a sinuous path around the planet at speeds which sometimes exceed 300 miles per hour. It passes unseen high above the horse latitudes – those breathless zones where ships often lay becalmed for weeks, 'standing mirrored in their own stagnant images'. Over the Sahara Desert, over the Sargasso Sea, over the sand dunes of Arabia and the bleak plateaux of Pakistan, streams this torrent of air. I almost said 'roars', but jet streams do not roar. Unlike the violent winds on earth that shriek around tall buildings, and clatter against barn doors, and whistle down chimneys, and moan at every chink of an open window, jet winds make no sound to betray their presence. No sound, at least, until some object intercepts their path, because, of course, it is not the winds themselves that we hear on earth but the vibration of their contact with all the objects in their path. Jet winds have nothing to rub shoulders with in the tropopause and so they pass as silent and swift as ghosts across the sky.

One of the early balloonists, Monck Mason, described the 'awful silence' experienced in a balloon carried along at the wind's speed in frictionless flight:

> The greatest storm that ever racked the face of nature, is in respect of its influence upon this condition of the balloon, as utterly powerless and inefficient as the most unruffled calm... [A man might ride high above a hurricane] traversing the skies at a time when one of the wildest and fiercest was exercising its utmost powers of devastation, looking down from his air-borne car and beholding houses levelled, trees uprooted, rocks translated from their stony beds and hurled into the sea . . . and all the various signs of desolation by which its merciless path is marked; he might nevertheless hold in his hand a lighted taper without extinguishing the flame, or even indicating by its inclination to one side or the other the direction of the mighty agent by which such awful ravages had been created.[5]

So might a man travelling in a balloon along a jet stream look down and see the cyclonic storm generated by a low-pressure area in the waves which he was calmly riding, see the boiling clouds and the driving rain, while he himself hung suspended in the quietness and startling clarity at the edge of the stratosphere.

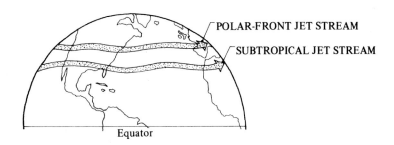

POLAR-FRONT JET STREAM

SUBTROPICAL JET STREAM

Equator

Fig. 7. Typical positions of jet streams in the Northern Hemisphere in winter.

The Southern Hemisphere has its polar-front jet and sub-tropical jet, too, and they follow tracks similar to the paths of their Northern Hemisphere counterparts. These four jet streams, together with their intervening bands of turbulent lower-atmosphere westerlies, can be visualized as two giant cyclones around the earth. If these winds could be made visible by means of a magic dye, an astronaut approaching the planet from outer space and looking down on the North Pole would see a swirling vortex of wind rotating counter-clockwise with the white Arctic Circle lying in the calm centre of the whirlwind. Viewed from just above the South Pole, the corresponding circumpolar vortex in the Southern Hemisphere would be moving clockwise, framing Antarctica.

Under certain conditions the polar-front and the sub-tropical jet streams approach each other and join together into one very powerful current with maximum wind velocities which generate unusually violent storms at ground level. It is possible also that jet winds help to spawn tornadoes when they slice across the tops of towering thunderheads, carrying the anvil eastwards and setting up small low-pressure eddies on the lee side.[6]

Because the energy that drives the jet streams comes from the temperature differences in the interacting air masses, changes in this temperature gradient affect the speed and the position of the streams. In the hemisphere that is experiencing winter, the contrasts are very great. The temperature at the pole ranges around $-40°F$ ($-40°C$) and at the equator about $85°F$ ($29°C$). During this season another jet stream develops near the pole that is shrouded in darkness. Known as the *polar-night jet*, it flows east through middle to high latitudes at heights of twenty to thirty miles.

In the summer, when the temperature at the pole hovers around $32°F$ ($0°C$), the temperature gradients across the hemisphere are much smaller and the jet streams become weaker. In fact, the polar-night and the sub-tropical jet streams disappear entirely. The absence of the sub-tropical jet in the Northern Hemisphere allows another important high wind current to become established. Aided by the seasonal flow of cool moist ocean air over the hot land, an *easterly jet* forms across southern Asia and Africa, bringing with it the characteristic monsoon rains. This jet blows in the lower stratosphere above the ten-mile level. A nonconforming member of the jet-stream family, it is the only easterly jet and it has no counterpart in the Southern Hemisphere.

The exact timing of the change from the domination of the westerly sub-tropical jet to the easterly jet is very critical to the agriculture in these lands of marginal rainfall. A delay in the reversal of the wind pattern can cause crop failures that result in the starvation of millions of people. A southerly shift in the path of this jet moves the belt of monsoon rains out of northern India and the sub-Sahara. This shift can be caused by a very small increase in the temperature contrast between the arctic and the tropics.

Knowledge about these remarkable rivers of air is an invaluable aid in making more reliable weather predictions and in monitoring the long-term changes that are occurring in the planet's weather. The position and the speed of the streams (especially the polar-front jet) help to foretell the location and the violence of the storms that will occur on earth. A plot of their movement with the seasons provides an excellent way of taking the world's temperature because the position of the jet streams is so closely related to alterations in heat flow between the tropics and the poles.

12. A band of cirrus clouds crossing the Nile Valley and the Red Sea marks the core of a jet stream.

Travelling silent and invisible through relatively narrow tubes of space in winding patterns around the planet, these high-flying currents of air might be able to hide in the sky were it not for the extensive weather data collection system now in operation. Information is constantly being collected by aircraft, by weather balloons, and by satellites that circle the earth far above the jet streams. Satellites that occupy an approximately circular orbit at an altitude of 22,000 miles make a complete revolution in twenty-four hours; thus their movement is synchronized with the rotation of the earth and they appear to remain fixed in the sky over one spot on the globe. Such satellites are able continuously to monitor at least one fourth of the earth. Other weather satellites travel in polar orbits. As the earth rotates beneath them they gradually scan the whole planet.[7] Photographs from these satellites are radioed back to earth. In these pictures jet streams occasionally betray their presence by a thin line of cirrus clouds, like the trail of crumbs dropped by Hansel and Gretel in the forest. [See Plate 12.]

At noon and midnight Greenwich Mean Time, approximately 700 weather stations around the world launch radiosonde balloons containing miniature weather stations which transmit information on the upper air as they rise towards the top of the troposphere. Carrying radar reflectors, they can be tracked even through clouds to yield measurements of temperature and wind speed and direction. All of this data is relayed by the Global Telecommunications Systems around the world. Together with the information from satellites and ground-level observations from about 9,000 other weather stations, 6,000 merchant ships, and hundreds of aircraft, these facts are fed into giant computers. Although there are some regions, such as the southern oceans, that are not thoroughly covered by this global weather watch, it does provide an impressive amount of information almost continuously, thereby enabling meteorologists to plot the wind-flow patterns of the earth. They can usually pinpoint the position and estimate the strength of even those elusive jet winds.

Just a few decades ago men did not dream that rivers of air were passing overhead at such fantastic speeds. Now as I look up at the intense blue 'emptiness' of a clear October sky I wonder how many other secrets still lie hidden in that innocent blue eye of heaven.

CHAPTER NINE

THE WHIRLWINDS

The voice of the sea is never one voice, but a tumult of many voices – voices of drowned men, the muttering of multitudes of the dead, the murmuring of innumerable ghosts, all rising to rage against the living at the great witch-call of storms.

– Old Breton proverb, as phrased by Lafcadio Hearn[1]

The Sea Islands of Georgia are not really islands at all, but low-lying peninsulas, hardly more than sandspits. The sea has crept inland in many places, making an intricate network of sinuous waterways and salt marshes where yellow sedge grass grows. In 1893 these islands were inhabited almost entirely by Negroes, freed slaves whose parents had been bought to work the cotton fields under white owners. After the Civil War they had stayed on in the islands, continuing to grow cotton and to live the only life they knew. Those who lived on the creeks and estuaries used them as highways to town, their boats flitting silently in and out of the tall marsh grasses. On higher ground two-wheeled donkey carts carried supplies and brought the cotton harvest to market. But there were no regular channels of communication with the mainland; so on 27 August the inhabitants of these islands did not know that a major storm had been sighted in the Caribbean heading westwards.

28 August dawned blustery with a breeze blowing out of the north-west, but soon it subsided to a few vagrant breaths. The day was suddenly oppressively hot; the sky was brazenly blue and bare of concealing cloud. Near one of the little communities that faced on the ocean, the men, women, and children were out in the fields picking cotton. Beyond the cluster of cabins they could see the broad grey face of the sea. It had a strange oily look to it that day; no little ripples or dancing movements broke the surface. The waves came in long low rollers breaking on the beach, slow and regular as a heartbeat. Far out on the south-eastern horizon a few clouds began to appear – thin lines that streamed out from one point like ribbons in a wind. By midday they had moved high over-

head; behind them a veil of clouds formed and grew until it covered a large portion of the south-eastern sky. The sun shone through pale and milky with a halo the colour of dust.

It was late in the day when the clouds finally obscured the sun and the cotton pickers looked up from their work to gaze apprehensively at the sky. Several of the old folks knew the signs – they had seen them before. The evil spirit was boiling up out of the south! Scattering like bright leaves before the rising wind, the people hurried back to their cabins, lugging their half-filled baskets of cotton. Chickens and cats were herded into the flimsy houses. Windows were boarded shut. The sky now was filled with low scudding clouds and in the south loomed a heavy bank of copper-coloured clouds which grew ever darker and taller. Clusters of people stood watching with helpless fascination as it came nearer and nearer, turning first dark grey and then black like an enormous cloak obscuring half the heavens. As it moved it seemed to close down upon them from above. The winds were still light and gusty, but the pounding of the surf on the shore had grown louder with a slow powerful boom like the beating of a native drum, four or five beats a minute.[2]

All at once the main body of the wind struck and the people rushed to take cover. With staggering impact the wind and the rain smashed into the low-lying land. It came in waves of increasing violence as successive banks of clouds marched overhead. The tall palms and scrub pines bent double before its fury. Doors and porches were torn loose from the frail houses and rain poured down upon them like a waterfall out of the sky. The waves lashing on the shores grew to mountainous heights. They broke over the low sand bluffs and lapped around the foundations of the houses. Families crouching together in their cabins could hear water swirling beneath the floor. Houses moved, swayed uncertainly, and broke apart. Walls fell away and the wind roared in, sweeping the inhabitants out of the cabins as a broom sweeps out crumbs. Babies, old people, and invalids were tossed into the turbulent waves that were breaking all the way across the land. Wherever a wall was still standing men and women crouched on the lee side, clinging together, crying and praying. 'We stood there the live-long night,' reported one survivor, 'like a passel of turkeys with their heads down.'[3] Children half waded, half swam to the nearest trees and climbed high into their branches. One family lashed themselves together around the trunk of a palm tree. On all sides trees were falling. One giant live oak hurtled through the air and landed flat in the water, its long festoons of Spanish moss streaming out on the waves.

Then suddenly the sea drew back as though the earth were taking a deep breath, and after a long, terrifying pause, it returned in a huge wall of water that swept towards the harbour. Funnelled by the encircling peninsulas of land, it grew higher as it advanced – a monster rising from the deep. A wave fifteen feet high fell on the village like a piece of the ocean dropped from the sky. As the water rushed back to the sea it carried with it hundreds of bodies, the carcasses of chickens, pieces of pink-and-lavender-painted wall, a rocking chair, and a broken tree with a family lashed to its trunk.

Days later this flotsam of the storm was washed ashore on the mainland of Georgia, mute evidence of the disaster that had struck the Sea Islands. When help finally arrived the rescuers found a dazed remnant of the population. Windrows of dead still lay upon the beaches and unidentified bodies drifted throughout the winding waterways. Hundreds of tons of dark sedge grass were strewn on the cotton and sweet potato fields. At least nine hundred lives had been lost in this storm, all but twenty-five of them blacks.

Many of the victims survived less than half of the hurricane. They had not experienced the strange eerie calmness of the eye, the centre of the storm, where the winds are still. Joseph Conrad in *Typhoon* described what it was like to experience the eye of a tropical cyclone in a ship at sea:

> ... the *Nan-Shan* wallowed heavily within a cistern of circular form in the depth of the clouds resting on the sea. This ring of dense vapours gyrating madly around the calm of the centre encompassed the ship like a motionless and unbroken wall of a blackness inconceivably sinister. Within, the sea, as if agitated by an internal commotion, leaped in peaked mounds that jostled each other, slapping heavily against the ship, and a low moaning sound – the infinite plaint of the storm's fury – came from beyond the limits of the menacing calm ... The ship was cut off from the peace of the earth. The wall rose high, with smoky drifts issuing from the inky edge that frowned upon the ship under the patch of glittering sky. The stars, too, seemed to look at her intently, as if for the last time, and the cluster of their splendour sat like a diadem on a lowering brow.[4]

The eye of a tropical cyclone is a strange enclosed environment like no other place on the face of the earth. The air is heavy, sultry, and hot – often as much as 32°F (18°C) above the air of the outer storm. The sun and the stars shine through a fine haze with a thin unreal light. The barometric pressure is extremely low; 26·18 inches – the lowest officially accepted reading – was recorded during a tropical cyclone in 1927 in the Philippines. (Some people have reported that such very low pressures strained their eardrums and brought the taste of blood to their mouths.)[5]

The sea at the centre of the vortex is wildly disturbed, foaming like the mouth of a mad dog. Great swells and breakers, running into the eye from all directions, clash and throw up tall plumes of spray. But above the torn surface of the sea, the air is calm. Many battered birds have been seen in the core of hurricanes, wheeling and circling in the fifteen- or twenty-mile radius which is sheltered from the fury of the winds.

The very low pressure in the eye causes the sea surface to be sucked up into a hill of water as much as ten feet above the normal level of the sea. As the storm moves, the hill of water moves also. Lashed furiously by the winds, heavy swells build up and run out ahead of the storm itself. These swells follow each other in a slow rhythm – four or five a minute – whereas the normal rhythm of waves that build up over a wide sweep of ocean is seven or eight a minute. The powerful, deliberate boom of hurricane waves brings the message of impending disaster to lonely islets and populous harbours hundreds of miles away.

As the centre of the storm moves across any land in its path, the hill of water crashes over the beaches like a tidal wave, inundating everything that lies within fifteen or twenty feet (four and a half to six metres) of normal sea level. High tide adds to the level of the storm surge waves, making them more dangerous, and V-shaped land formations that funnel the waves magnify the devastating effects. The Bay of Bengal where the Ganges River enters the sea has been the scene of several of the most terrible typhoon floods. On the seventh of October in 1737 a typhoon slammed into the bay and a surge wave forty feet high struck the harbour. It is said to have destroyed 20,000 ships and drowned 200,000 people, flooding more than 8,000 square miles of the Ganges Delta. In 1970 another disastrous typhoon swept in from the bay across the low-lying islands of Bangladesh. It is estimated that nearly 300,000 died in this storm.[6]

The eye of the tropical cyclone is the essential core of its being. Condensation of water in the hot rising air around the perimeter of the eye continuously provides the energy that drives the storm. Hurricanes form in the doldrums, out over the tropical seas, in late summer or early fall when the water temperature has reached a maximum of close to 80°F (78°F is a necessary minimum for the birth of a hurricane). At this time also the doldrum belts have moved far enough north so that the Coriolis deflection produced by the earth's rotation causes the column of air to turn in a giant whirl as it rises. On the equator itself there is no Coriolis effect. Therefore, the vortex of the tropical cyclone cannot become established in a current of hot air rising near the equator. In both directions away from the equator, the Coriolis effect becomes more and more pronounced. The breeding place of most tropical cyclones is at 6 to 30 degrees north latitude during the season of the year when the water there is warmest.

Exactly why some columns of air begin to funnel upwards rapidly enough to give birth to the tropical cyclone is a matter of debate among meteorologists. Some postulate hot spots formed by the ocean currents where the water surface is distinctly warmer than the surrounding water. Others believe that waves of low pressure in the upper atmosphere cause the column of air below it to be drawn upwards as hot air is drawn up a chimney. These explanations are not mutually exclusive. In fact, the most generally accepted view today is that hurricanes are generated when these two conditions occur together and reinforce each other. [See Plate 13.]

Given a column of rising air with the proper whirl imparted to it, the formation of a hurricane is easy to understand. The hot air off the ocean is humid, and as it spirals upwards it is cooled by expansion to the point where it can no longer hold so much moisture. Clouds and raindrops form, with the release of large quantities of heat energy, which warms the air again and reinforces the updraft. The spiralling winds increase in speed; more moisture-laden air rushes into the base of the spiral to equalize the low pressure caused by the updraft, and the updraft grows stronger as the water condenses out into torrents of rain.

A partial vacuum is caused by the expanding updraft and by the centrifugal force of the spiral motion, which sends the raindrops spinning to the edge of the vortex just as water swirling down a drain-pipe is thrown to the periphery of the little whirlpool, leaving the centre relatively empty. The partial vacuum causes some air to be drawn down into the centre of the core from the upper atmosphere. This air, warmed by compression as it descends, helps to maintain the pool of hot atmosphere in the centre of the tropical cyclone. The greater the temperature difference between the eye and the surrounding air, the more violent the storm.

13. Hurricane Gladys as seen by the astronauts of Apollo 7 in October 1968.

A hurricane, or a typhoon as it is called in the Pacific, is like a thunder-storm multiplied many times and given a strong spiral spin. It strikes with the force equal to the simultaneous explosion of 1,000 atomic bombs like the one that levelled Hiroshima.[7] A hurricane can lift two billion tons of water a day, and this enormous load is emptied out of the storm as rain. It is literally true that the air picks up a piece of the sea and drops it again miles away.

Since the energy that drives the tropical cyclone comes from the con-densation of water vapour, the storm must have vast quantities of hot humid air to sustain it. As it passes over a large body of land and inhales drier air, it loses some of its power and gradually fizzles out. Storms that move out over colder water also lose their strength. The path that a hurricane takes over the sea is not easy to predict. Its course is generally a curve to the west or north-west which then swings north or north-east. But it is often erratic. Some storms move south or east, farther out into the Atlantic. A storm that has been progressing east can suddenly turn around and head west. For this reason it is very important that these storms be watched almost hourly and the threatened coastal areas warned, giving the people time to protect themselves and their property against the storm.

During the Second World War a very fine system for hurricane watch-ing was set up by the U.S. Navy, Air Force, and Weather Bureau. Ships at sea encountering torrential rains or rapidly falling barometers notified the Weather Bureau, and these ominous findings were immediately sent to the military bases, where planes were always ready to take off on a hurricane hunt, carrying airborne weather stations into the heart of the storm. Planes from the Philippines patrolled the China and Philippine seas; planes from Guam watched the regions east of the Philippines. The Gulf of Mexico and the Caribbean were covered by planes based in Florida and Bermuda.

When a potential storm was sighted over the Caribbean area, for example, the U.S. Navy dispatched a reconnaissance plane from its base in Florida in the morning; an Air Force plane from Bermuda took up the watch in the afternoon. The information collected by these flights was radioed to the Joint Hurricane Warning Center in Miami, where Weather Bureau scientists analysed it and predicted the course of the storm. Radio warnings went out to the threatened communities. In the

Pacific area the data was collected and processed at the military head-quarters in Guam.[8]

Navy planes that hunted the hurricanes and typhoons carried eight to ten men: the pilot and co-pilot, the navigator and engineer, radio and radar operators, a weather observer, and men to monitor the special electronic measuring devices. Flying towards the storm they usually maintained an altitude of 1,000 to 1,500 feet. As they approached the cyclone they circumnavigated it, flying counterclockwise in order to travel with the winds. In the Northern Hemisphere the right half of the hurricane as you look along the path of its motion is the most violent portion because in that region the speed of its forward motion is added to the speed of the whirlwinds. On the left side these motions cancel each other and the violence is slightly moderated. The reconnaissance planes sought out this 'back door' and flew through it towards the centre of the hurricane. As they neared the vortex they descended to about 500 feet in order to fly under the densest clouds and keep the torn surface of the sea in sight. Radiosondes were dropped onto the ocean in the eye of the hurricane. These small balloon-borne transmitters send out radio signals giving temperature, humidity, and barometric pressure. Blind flying conditions usually prevail near the hurricane core, so pilots tried to fly in under the 'soup' and take drift-meter readings of the waves to aid in calculation of the wind speeds. The extreme turbulence just at the edge of the vortex often made veteran pilots seasick. Wind speeds there sometimes exceeded 200 miles per hour.

Reconnaissance planes are still used in the hurricane warning system. However, in the years since it was set up in 1944, technology has made available many sophisticated new instruments which help in the effort to identify and predict tropical cyclones. Radar, invented during the war and used even on the earliest reconnaissance flights, has been continuously improved. It is an invaluable aid to the hurricane hunters, enabling them to 'see' the eye of the storm when they are still a long way from it. The eye shows up on the radar screen as a luminous crescent made by rain bands around the centre. Land-based radar identifies storms that approach within a hundred miles of the coast.

However, it is the advent of weather satellites that has really revolutionized the hurricane warning service. Several such satellites are now constantly monitoring the globe. They pick up weather disturbances as they form out over the oceans, measure the speed and circulation of the winds, and follow the course of the storm. Sensitive receivers in the satellites also measure the heat radiated by each portion of the earth's surface. These infra-red photographs provide detailed information on the temperature of the sea and can detect the hot spots that cause the air columns to start rising. Thus, potential hurricanes can be identified before they become dangerous storms. From infra-red pictures, cloud heights and movements can also be monitored and wind speeds can be inferred.

These data, transmitted back to earth every thirty minutes, are processed by electronic computer systems at the National Hurricane Center in Miami and at the Joint Typhoon Warning Center in Guam. When a potential storm is identified reconnaissance flights are sent out to monitor the disturbance. The U.S. Air Force now flies all of the routine hurricane watch flights from bases near Biloxi, Mississippi, and on Guam. The planes fly into the storm at higher altitudes than the earlier flights – usually about 10,000 feet. From this safer height they are still able to drop radiosondes accurately into the eye and around the perimeter of the storm. These instruments, perfected over the past few decades, radio back information on temperature, pressure, and humidity. Wind speeds are measured by radar. The data collected from these flights, from the satellite photographs, and from land-based radar provide such remarkable coverage that a major storm cannot approach the coast of the United States undetected. Before an adequate watch system was devised storms like the one that hit the Sea Islands took a fearful toll of life every year. In recent years the loss of life has been small even though the storms themselves rage with the same fury that they always did.

Fascinating new facts about hurricanes have also been revealed by the weather satellites. The ocean just west of the coast of Africa has been discovered to be a major breeding ground for tropical storms. Hot air off the Sahara Desert meeting the cooler ocean air appears to provide the atmospheric conditions that generate approximately half the hurricanes in the Atlantic. Before the advent of satellites it had been assumed that the majority of hurricanes originated in the Caribbean. From satellite pictures it has also been discovered that about twice as many tropical storms form off the west coast of Mexico as had been supposed.[9]

All of these facts are adding to the understanding of the birth and life cycle of a tropical cyclone and are speeding the day when hurricanes will not only be predicted but controlled. The economic incentive for modifying the impact of tropical storms is very large, for although the accurate prediction and warning system has dramatically reduced the death toll from these storms, the property damage has sharply increased as population has grown and more expensive construction has gone up along the shores. In 1970 a single hurricane, Celia, caused $454 million worth of damage, more than the damage caused by all the hurricanes during the five-year period from 1925 to 1929.[10]

Many ideas have been proposed for taming hurricanes, but the one which is believed to offer the best potential is seeding with silver iodide crystals. Experiments with several hurricanes over the past few years have shown that the power of the storm can be reduced by seeding the clouds in the area just outside the wall of the eye. The seeding causes more rapid precipitation in this region, liberating latent heat and warming the clouds surrounding the eye. Thus the temperature difference between the warm eye and the cooler surrounding clouds is reduced and the balance of the forces that drive the storm is disturbed. In 1969 Hurricane Debbie was seeded in this way and the wind speeds dropped by 10 to 15 per cent. This encouraging result raised hopes of routine modification of tropical cyclones. But there are hazards, both legal and ecological, which must be considered before any large-scale programme of this type can be safely undertaken.

The dynamics of a hurricane are still not perfectly understood and no two are exactly alike. Each storm is a law unto itself. There has been some fear that seeding could cause a change in the direction of forward motion, making the storm descend upon shores that might otherwise have been spared. If a storm is very destructive even after it has been seeded, the victims might maintain that seeding had increased rather than reduced the damage. And finally the possibility of harmful side effects must be considered. Will the introduction of tons of silver iodide into the atmosphere cause ecological damage that has not yet been anticipated?[11]

However, scientists working on a hurricane modification study called Project Stormfury are optimistic that the detrimental side effects will be negligible and that with improved theoretical understanding of the physics of hurricane clouds these greatest of storms on earth can be tamed.

The hurricane is beginning to yield its secrets and a large part of the information which has made this possible has been collected by the daring pilots who have steered their fragile aircraft in and out of the heart of the storm, taking its temperature, testing its muscle, and feeling the pulse of that gyrating ring of low pressure which pumps energy from the hot sea to power the life and vitality of the storm. But detailed information of this kind is not available for the tornado; this vicious little cousin of the hurricane does not give up its secrets so easily. Even the most daring pilot would not fly his plane into the heart of a twister. The few that have accidentally blundered into one have been torn wing from wing, the pieces spat out after they had been thoroughly masticated in the tornado's ugly maw. Only a small select number of human beings have looked into the funnel of a tornado and lived to describe it.

One of the most famous accounts was that given by a farmer in Kansas. A thunderstorm had swept across his property that afternoon, bringing with it heavy hail and driving winds. When the storm abated Will Keller and his family rushed outdoors to see what damage had been done to their wheat crop. Just a few hours before, the beautiful field of young wheat had stretched in a deep-piled carpet of green. Now the crop was beaten and matted in enormous swirls that looked like giant hen's nests laid one beside another. As Keller and his family stood looking at this scene of devastation the sky began to grow dark again and an eerie sickly-green glow intensified the colour of the beaten field. Keller glanced up and was appalled to see three long funnels dangling from the greenish-black base of an umbrella-shaped cloud. One of them seemed to be heading straight for the Keller farm. The cyclone cellar was not far away and Keller rushed his family into it. As he was closing the trapdoor he paused to take one more look at the storm. He had seen a number of tornadoes and did not feel panic-stricken by them. He knew they moved slowly and there would be plenty of time for him to slam the cellar door shut as it approached. Furthermore, he was curious to see what a funnel cloud looked like really close by. So Keller stood watching the three tornadoes sweep towards him across the Kansas plains and later he reported in detail what he saw:

> Two of the tornadoes were at some distance away and looked to me like great ropes dangling from the clouds; but the near one was shaped like a funnel with ragged clouds surrounding it. It appeared to be much

14. Tornado near Dallas, Texas, 1957.

larger and more energetic than the others, and it occupied the central position of the cloud, the great cumulus dome being directly over it.

As I paused to look I saw the lower end which had been sweeping the ground was beginning to rise. I knew what that meant, so I kept my position. I knew that I was comparatively safe and I knew that if the tornado again dipped I could drop down and close the door before any harm could be done.

Steadily the tornado came on, the end gradually rising above the ground. I could have stood there only a few seconds, but so impressed was I with what was going on that it seemed a long time. At last the great shaggy end of the funnel hung directly overhead. Everything was as still as death. There was a strong gassy odor, and it seemed that I could not breathe. There was a screaming, hissing sound coming directly from the end of the funnel. I looked up and to my astonishment I saw right up into the heart of the tornado. There was a circular opening in the center of the funnel, about fifty or one hundred feet in diameter, and extending straight upward for a distance of at least one half mile, as best I could judge under the circumstances. The walls of this opening were of rotating clouds and the whole was made brilliantly visible by constant flashes of lightning which zigzagged from side to side. Had it not been for the lightning I could not have seen the opening, not any distance up into it anyway.

Around the lower rim of the great vortex small tornadoes were constantly forming and breaking away. These looked like tails as they writhed their way around the end of the funnel. It was these that made the hissing noise . . .

The opening [at the upper end] was entirely hollow except for something which I could not exactly make out, but suppose that it was a detached wind cloud. This thing was in the center and was moving up and down . . .

After it passed my place it again dipped and struck and demolished the house and barn of a farmer by the name of Evans. The Evans family, like ourselves, had been out looking over their hailed-out wheat and saw the tornado coming. Not having time to reach their cellar they took refuge under a small bluff that faced to the leeward of the approaching tornado. They lay down flat on the ground and caught hold of some plum bushes which fortunately grew within reach. As it was, they felt themselves lifted from the ground. Mr. Evans said that he could see the wreckage of his house, among it being the cook stove, going round and round over his head. The eldest child, a girl of seventeen, being the most exposed, had her clothing completely torn off. But none of the family was hurt.[12]

Another man who demonstrated amazing composure as well as a death-defying desire to record this awesome sight was Maurice Levy, a cameraman for the National Broadcasting Company. One April afternoon in 1957, he saw a funnel cloud heading straight for downtown Dallas. As the tornado approached he jumped in his car and raced towards the storm centre through streets choked with panic-stricken people.

'Suddenly I was surrounded by screaming,' Levy recalled, 'people were shouting, "Save us, save us for Jesus, mister." About a dozen of them managed to squeeze into my car. I tried to tell them I was headed toward the storm, not away from it, but this was so improbable they wouldn't believe me.'[13]

Navigating his carload of hysterical passengers through the chaos of the city streets, Levy drove close to the edge of the tornado and was able to keep pace with it for several minutes while he shot 600 frames of film footage looking up into the hungry mouth of the funnel. These photographs gave the Weather Bureau experts an unusual opportunity to study wind speed and pattern of flow inside a twister. Individual pieces of debris like walls and tabletops could be identified and followed through the whirling path of the vortex. Since the time lapse between frames of the motion picture was known, wind speeds within the funnel could be estimated. The results indicated that velocities were not as high as had previously been supposed. Two hundred miles per hour seemed to be the maximum. From this and similar estimates of other tornadoes scientists have concluded that the enormous impact of the tornado results not just from the winds themselves but also from the extremely low pressure in the funnel. Houses and other buildings often explode because of the pressure difference between the inside and the outside when the funnel passes over them. The barometric pressure of a large tornado is judged to be somewhere in the range of 25·5 inches.

Tornadoes come in many sizes. Statistical studies of the tornadoes which occurred in 1972 revealed a broad spectrum of intensities. The majority were small and short-lived, lasting only a few minutes. About 90 per cent had maximum wind speeds of 150 miles per hour; their path lengths were 10 miles or less and the width of the swathe they cut was less than 500 feet. However, about 5 to 10 per cent of the tornadoes were much more powerful storms, cutting paths of destruction up to a mile wide and 32 to 100 miles long. Wind speeds within these maxi-tornadoes have been estimated as high as 300 miles per hour.[14]

Although tornadoes have been observed in many countries and may occur at any time of year, by far the greatest number of really dangerous ones occur in the United States in the spring. They are spawned in violent thunderstorms where hot humid air encounters cold dry air masses. All thunderstorms produce strong updrafts and also downdrafts occurring very close to each other. Some combination of forces, still unexplained, causes a very intense localized effect which starts the formation of the tight spiral funnel. Some scientists postulate that severe electrical discharges can spark abrupt changes in air density and provide the enormous concentration of energy necessary to drive the whirlwind. Will Keller and other observers have reported brilliant lightning and an unnatural glow to the air in the immediate vicinity of a tornado. On the other hand, some tornadoes do not seem to exhibit unusual electrical activity.[15]

Meteorologists speculate that the formation of a tornado is triggered by sudden changes in air flow at both the top and bottom of the storm. Photographs taken from high-flying aircraft such as the U-2 planes show that many thunderheads which breed tornadoes reach to very high altitudes, breaking into the barrier of the tropopause and lifting it in a series of tall turrets overshooting the anvil top of the main cloud.[16] The tops of these turrets may grow and collapse very rapidly. Perhaps – one theory goes – a high wind blowing across these tops could cause sudden collapse and, in the lee of this wind, low-pressure eddies would be formed, starting the powerful updraft of a tornado.[17]

T. Theodore Fujita at the University of Chicago suggests that as the air ascends to these great heights, the temperature decreases rapidly and the air becomes heavy enough to fall, starting a downdraft towards the earth. In the volatile, unstable environment of the thunderstorm, a rapid downdraft like this could trigger the formation of a twister.

Radar scans of thunderstorms show that those which produce tornadoes often have a hook-shaped region where heavy rain or hail reflect

15. Twin water spouts in the Gulf of Mexico.

the radar signal. These hook echoes have diameters of several miles and often occur along the trailing right flank of severe storms which usually move from south-west to north-east. The tornadoes typically emerge from near the southern edge of the hook-shaped regions of these storms.[18]

Unfortunately, however, not all tornado-breeding storms carry this characteristic feature. Therefore, the radar signals cannot be used as a totally reliable indication of approaching tornadoes, but the sighting of such an echo in a thunderstorm usually leads to a tornado warning.

A study of the location and frequency of tornadoes in the United States turned up some long-term changes which have not been fully explained. The frequency of tornadoes reported in the United States since 1916 has increased dramatically. The average annual count during the first ten years (1916–25) was only 100. Thirty years later the annual count had increased to 200. By 1953 it had risen to 437 and since 1973 it has exceeded 700.[19] The most probable explanation of this apparent increase is greater efficiency of reporting and denser distribution of population. Since the majority of tornadoes have path lengths of only a few miles, in a sparsely inhabited region tornadoes of this size would go undetected. This explanation, however, cannot be put to any rigorous test, and so the possibility exists that a real increase has occurred.

An even more unexpected fact was discovered when scientists analysed tornado records going back to the early years of this century. Using average fatality figures as an approximate measure of tornado activity, they found that the geographical centre of the heaviest death-toll had moved in a roughly circular clockwise pattern. From 1926 to 1933 the centre was in northern Arkansas, Missouri, and Illinois. It shifted to the south-east states (Alabama and Georgia) in 1934–40, to Oklahoma in 1941–8, Kansas in 1949–55, Iowa in 1956–63, and back to Illinois in 1964–70, thus completing a circuit in forty-five years. The latest five-year average shows a move to the south-easterly direction again around the clockwise pattern. Is there any reason why the centre of maximum tornado activity should shift in a forty-five-year cycle? One of the guesses made by the scientists who conducted this research is that the cycle may be tied to sunspot activity, which oscillates on a period of slightly over eleven years. Four times this period makes forty-five. Is this relationship coincidental or meaningful? And how could sunspot activity influence the location of tornadoes? Here is another fascinating unsolved puzzle to challenge the imagination of the scientists who are trying to understand the origin of these killer storms.[20]

As the years go by, information continues to pour in about the shape, occurrence, and life cycle of tornadoes. Certain atmospheric and geographical conditions appear to influence their formation. They occur in many countries around the world, usually between 20 and 50 degrees in both hemispheres. In the United States, Europe, and India the maximum number occur in spring and early summer. Italy, Israel, and Japan, on the other hand, have active tornado seasons in autumn and early winter. In Australia and New Zealand there is no marked seasonal variation.[21]

Local terrain and temperature conditions may also affect the strength and path of these storms. The heat islands and surface friction created by cities, especially large ones with populations over four million, appear to dissipate or deflect tornadoes. Statistics gathered by Fujita show a very low incidence of tornadoes in Chicago and Tokyo. But there is a high incidence in a fringe belt around these cities.

Under one exceptional circumstance an outbreak of tornadoes occurred in Tokyo. An earthquake struck the city in 1923. Throughout the entire metropolitan area fires were burning out of control. The ascending smoke and hot air produced cumulonimbus clouds, some rainfall, and *tatsumaki*, or dragon whirls, as tornadoes are called in Japan. Most of these were small, but one was large enough to pick up a house with seven or eight people in it. One hundred and twenty *tatsumaki* and dust devils powerful enough to do some damage were reported.[22]

Dust devils are very closely related to tornadoes but they are not associated with a cloud. Their funnels are transparent, so that all the dust, sand, and other objects drawn into them can be clearly seen. Although most people think of dust devils as relatively minor disturbances, occasionally they are powerful enough to pick up an automobile, a feat which would classify them as comparable to an average-sized tornado. Whirlwinds that appear over water are usually called water spouts, but when associated with a thunderstorm they are a form of tornado. Sometimes they move out over land and cause considerable damage.

Satellites are constantly monitoring the thunderstorms that spin off tornadoes, but since the funnel clouds occupy a very small cross-section of space, photographs of higher definition are needed in order to study the characteristic patterns and physical variables that might give forewarning of these storms.

Perhaps someday we will understand how such a vicious vortex of motion can build up in the sky – a vortex that has the power to lift freight cars, to pull the blankets off a bed and suck them up a chimney, to pluck the feathers from a chicken, to strip trees of their greenery and bark – leaving just the pale shining trunks where drops of sap spring forth like blood on a skinned body. The fantastic strength of the twister is all the more remarkable when we consider that the roots of this violence lie concealed in a calm body of air nurtured in sunny, tranquil skies. The energy in the water vapour hidden there is quiescent and widely dispersed until – by some mysterious means – it becomes packed into a narrow funnel of space. To achieve such an extreme energy density it must be confined and driven by tremendous forces, just as the surge wave of the hurricane is funnelled into a towering wall of destruction by the outstretching arms of land that form a long narrow bay of the sea. But the banks of a V-shaped bay are solid, unyielding objects, obviously capable of confining and compacting. The forces that concentrate the energy in the tornado are built of volatile atmosphere and the random motion of its molecules. It might be reasonable to suppose that any chance concentration of heat energy would distribute itself in ever widening circles throughout the thinning molecules of the atmosphere up to the very farthest reaches of matter where the sky is deep violet and the stars can be seen in the day. Obviously such common-sense assumptions are too simplistic to describe the complex, dynamic, and delicately balanced medium that enfolds our planet – a medium that seems almost to have a temperament, a will, a vitality of its own.

CHAPTER TEN

LEE WAVES, WILLIWAWS, AND RIPPLES OF SAND

If you wish merely to listen *to the sky, or* smell *the sky, or* feel *the sky with your fingertips, do that, too.*

– Lawrence and Lee[1]

Plumes of ice crystals stream on high winds from the tallest mountain-tops; white veils of mist hang in the valleys; and in tranquil tropical waters volcanic islands lie crowned with wreaths of clouds. Everywhere around the world we see evidence of the intimate interaction between air and earth.

Across the Sahara and the desert peninsula of Arabia, sand lies in long seifs, blown by the wind into solid waves that corrugate the surface of the land. [See Plate 16.] And the burning sands warm the air that shapes them, creating rising currents that move out over the cool moist ocean breeding thunderstorms. Occasionally, by some chance rendezvous with just the right wind patterns aloft, these currents give birth to monsters, the hurricanes that descend with devastating force on lands thousands of miles away.

When the wind off the Sahara turns north and blows across the Medi-terranean, it is known as the sirocco, a wind which distributes millions of tons of reddish-coloured sand over Europe and adds a startling red colour to rains and snowfalls. In the year 582 a sirocco brought a 'shower of blood' to Paris and produced such fright among the inhabitants that they were said to have rent their 'blood-stained' garments and repented of their sins.[2]

When land has been stripped bare of its vegetation by drought and the plough, the topsoil is picked up by the winds and distributed around the globe. During the great American drought of the 1930s, farms in Okla-homa and Kansas were swept away and scattered as grains of dust across the whole east-central portion of the continent.

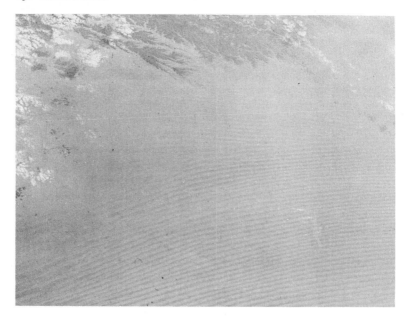

16. Sand is blown by the wind into seifs on the desert peninsula of Arabia.

Both matter and energy are constantly being exchanged between the air, the sea, and the land with its variegated surface, its tender skin of living things. The moan of the wind as it passes through a forest is the audible evidence of an interchange of energy between the air and the trees. The wind loses some of its strength by friction with innumerable pine needles, leaves, small sharp branches, and even the rough scaly bark of the trees. Experiments have demonstrated that after passing through only 100 feet of a mature forest, a wind loses 20 to 40 per cent of its original force. A passage of 400 feet reduces the velocity by 93 per cent.[3] This slowing of wind by a forest is easily perceived and measured, but the same effect occurs to a lesser extent whenever air passes across matter. If we were very small – say, as small as the grains of pollen that float in the summer breeze – we would observe that the surface of things, even those objects, such as rocks, that seem perfectly solid to us now, are actually quite porous, with billions of tiny interstices where the air can move in and out. As the wind passes over a rock some of the air blows through its

17. Bryce Canyon.

surface as though it were passing through a miniature forest. Drops of moisture condense out of the air and collect in the spaces. When they freeze they expand, causing the rocks to split. The little fragments are washed away by the next rain. Beaten and rolled by the elements, they eventually become grains of sand or fragments of soil. So throughout millions of years the atmosphere breaks up the rocks and tears down the mountains, creating the loose, friable surface in which life can take hold. The air and the rivers that flow from it carve out deep canyons, isolate towering buttes, and hollow caverns out of buried limestone. Some types of rock yield more rapidly than others to the eroding action of wind, rain, frost, and thaw; so the softer ones give way first, leaving exposed layers and strange skeletons of harder rock – the pink and white granite seams in the Grand Canyon, the delicately shaped limestone spires of Bryce Canyon, the rust and ochre sculpture of the Badlands, and the fantastic brown cones and chimneys of Cappadocia.

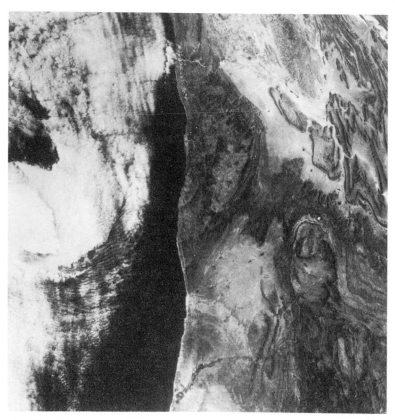

18. Wind deflected by tall mountain peaks has created this clockwise stratocumulus cloud formation off the coast of Morocco.

Not only does the atmosphere sculpt and mould the land, but the surfaces of the earth shape and influence the weather. As the air flows across matter, encountering the myriad textures and obstructions, it is deflected, bubbled, swirled, and roiled into an intricately complex pattern of motion, like a mountain stream pouring downhill over rocks and pebbles. [See Plate 18.]

The interaction of the air and the earth would be much simpler if the earth were uniformly covered with water. The smooth uninterrupted surface of the sea would cause only very minor disturbances in the air flow. Then prevailing winds would blow as reliably as the trade winds do now. Predictable weather would occur in seasonal cycles and the extremes of weather would be very much reduced because water does not warm up or cool off as rapidly as land. But the presence of land masses, the assorted sizes and shapes of the continents, the different thermal characteristics of land and sea create a dynamic flow of great complexity. Hundreds of special localized climatic conditions exist around the globe. These distinctive weather patterns are such important aspects of life in each locality that they have been given folk names and have even acquired a status approaching that of local gods. Guy Murchie collected 127 names of regional winds that varied all the way from the fierce 'landlash' of Scotland to the gentle breeze known as 'a sigh in the sky of China'.[4]

The most extreme weather conditions are not necessarily found at the poles or the equator. They occur on large land masses in the regions that experience the least modifying effect of winds off the sea. Land – especially bare rock and sand – responds very rapidly to radiant energy from the sun. Less energy is required to raise the temperature of soil than to raise the temperature of water a comparable amount. Forests and fields covered by dense vegetation have thermal characteristics that lie between those of water and bare soil. At night when the sun goes down the land cools off more rapidly than the lakes and oceans. Even in the hottest seasons in the desert when the daytime temperature may approach 130°F (54°C), the temperature at night drops off to 50°F (10°C) or even down to freezing. In general, the lands that lie farthest downwind of the oceans experience the greatest temperature differences between day and night, winter and summer.

Severe weather also occurs at high altitudes, where the air is thin and the wind's force is untamed by contact with city, forest, and field. In Tibet at 16,000 feet the inhabitants have become acclimatized to winds that blow constantly at gale force. By mid-morning, when it becomes impossible for them to make themselves heard above the roar of the wind, the natives bury their faces in their coat collars and go silently about their business. Mount Washington in New Hampshire is only a little more than 6,000 feet high, but the weather observatory on the top of the mountain has recorded wind-gust velocities as high as 230 miles an hour – the highest ever officially measured on earth.[5]

In the Southern Hemisphere the continent of Antarctica, which coincides with the geographical South Pole, is a forbidding place where winds blow with relentless violence. Day after day, month after month, wind speeds measured in Adélie Land averaged 50 miles an hour. However, in the Northern Hemisphere seas occupy most of the Arctic Circle and exert a modifying influence. The coldest lands lie hundreds of miles farther south – Siberia, the Dakotas, Greenland, and the plains of western Canada. Verkhoyansk in Siberia is reputed to be the coldest town in the world. In January temperatures there range from $-13°F$ to $-80°F$ $(-25°C$ to $-62°C)$. If the air is still and dry, exposure to these temperatures is not life-threatening, but when the winds blow out of the north at 50 or 60 miles an hour, carrying a blinding swirl of snow and ice crystals, the conditions outdoors are unendurable. Natives of Siberia call such winds by many names – the Chief, the Old Man, the purga. I. W. Shklovsky, a Russian writer, described his introduction to the Chief when he was visiting a Yakut tribe at Sukharnoe:

> After the first gust the wind kept up a continuous, monotonous scream, like the sound of escaping steam from a million engines, dreadful to hear, and the hut shook violently. The moss plugging the cracks in the log walls was cut out by the wind as though with a chisel, and its icy breath whistled through the hut. I wanted to know what it was like outside. 'Don't go out,' said my host, ' "the Chief" will not like it.'
>
> But, seeing that I was determined to go out, they fastened a long strap round my waist, holding the end firmly in their hands. Hardly had I opened the door when I was flung violently to the ground. The hard snow, congealed by sixty degrees of frost, was dug out by the fury of the wind, and the air was full of ice crystals, which whipped my face like molten metal, and in an instant my cheekbones were frozen. I was as one blind in the impenetrable darkness which surrounded me, and the only sound to be heard was the insistent howling of the wind, which dominated all else.
>
> Where could I go? Where was the door? Had it not been for the strap held so firmly by my friends, it would have gone hard with me, for I could not possibly have found the door; but I managed to crawl on all fours to the threshold. 'Well, friend, what did "the Chief" say to you?' my hosts asked jestingly, when, half frozen, I crawled into the warmth of the hut.[6]

. . .

Nomad bands, Eskimos, and other men caught out in savage storms like this can save themselves only by stretching out on the ground, covering themselves and their animals with blankets and pelts. In this way they are sometimes able to survive the storm.

At the other extreme, in the searing heat of Arabia, on the Sahara Desert, on the bleak plateaux of Turkestan and Iran, windstorms are like a burning breath from the mouth of hell. These winds have many names. The tebbad, the 'fever wind', sweeps across Turkestan; the simoom is the 'poison wind' of Arabia and Libya; the harmattan is the parching wind of Algeria and Morocco; the haboob is the scourge of the Sudan, as the khamsin is of Egypt. These are the withering winds baked in the oven of the desert, laced with sand, dust, and even pebbles swept up from the stony floors of the hammadas. They act like huge sand-blasting machines, abrading everything they touch.[7] The storms can be seen approaching across the flat wastes of the sand dunes and the tundras – a dark grey or yellowish cloud suddenly obscures the sun. When caravans travelling across the desert see these signs the men immediately dismount from the camels. The animals lie down on the sand with their hindquarters towards the storm and stretch their long necks out on the ground. The men crouch low behind the beasts, drawing their flowing robes and turbans close about their bodies, taking special care to cover their heads. On all sides of them sand moves and flows as though it were water. It piles over and around them, sometimes almost burying the caravan.[8]

Fortunately these dangerous climatic extremes do not affect most of the places on earth. Nearly 80 per cent of the land areas lie near enough to a large body of water to enjoy the cleansing and modifying effects of sea breezes.[9] Along the shores of the oceans and the large lakes, during the summer the winds follow a typical diurnal pattern. In the early morning the air is calm. The surface of the water is smooth as glass; only tiny water movements – hardly big enough to be called waves – lick the shore. Then, as the sun rises higher in the sky, the land becomes hotter than the water. The air over the land rises and the sea air flows in to replace it.

The sea breeze comes shyly at first, making light dancing footprints on the water. Then it retreats and the surface becomes like glass again. In a few moments the ripples reappear somewhere else, moving swiftly towards the land. Now the whole surface of the sea is sparkled with swirling ripples. The breeze flows steadily onshore, growing stronger until mid-afternoon. It brings refreshing coolness to the hot land – the taste of salt, the clean, tangy smell of brine and crustaceous sea creatures and kelp drying on the sand. As the sun swings lower in the sky the breeze gradually drops off. At night the land cools; the air settles down heavy again upon it. The air flow is reversed and the breeze returns to the sea carrying with it a strange mixed burden of dust and dandelion seeds, of pollen from pine forests and soot from steel mills. The hot acrid smell of city streets is mingled with the fragrance of ploughed fields and new-cut lawns.

In some parts of the world land and sea winds alternate on a seasonal basis in just as reliable a rhythm as the diurnal pulse caused by sunlight and darkness. Where temperatures day after day build up unusually high in the summer – as, for example, in south-eastern Asia, Arabia, and Australia – the rising hot air causes low-pressure areas that extend for millions of square miles. Cool moisture-laden air rushes in from the seas, and these atmospheric-pressure differences are so great that they override any small day–night variations. The sea wind dominates for a whole season. In winter the land cools off and the process is reversed. These winds that change direction 180 degrees from winter to summer are called monsoons; the name comes from the Arabic word *mausim*, meaning season.

The position of mountain ranges helps to set prevailing wind patterns. Very high ranges like the Himalayas prevent even the most vigorous cold fronts from sweeping far south into India, Burma, and Indochina. Italy is protected by the Alps, and Spain by the Pyrenees. In the United States, on the other hand, there is no east–west mountain range to block the southerly flow of cold northern winds; so freezing weather does occasionally penetrate into Texas, Florida, and the other Gulf states. When a norther descends into Texas the temperature often drops as much as 25°F (14°C) in one hour. As this wind crosses the Gulf of Mexico and strikes the coast of Central America, it is known as a 'norte' and brings a spell of unpleasant weather that may last for a week or more.

Deep valleys between mountains act as troughs to channel the winds and magnify their effects. Roaring through mountain passes and down river valleys, bleak cold winds bring chilling weather to lands that normally enjoy a gentle climate. The mistral, that dreaded scourge of southern France, is a north wind that is funnelled through the valley of the Rhône between the Alps and the Cévennes Mountains. Descending on Provence, it sweeps everything before it. Slater Brown describes the mistral, a perennial but unwelcome visitor:

> It has the ill-natured habit of scattering roofing tiles about, knocking down chimneys, blowing small children into canals, tumbling walls onto the unsuspecting natives, upsetting wagon-loads of hay and sometimes overturning freight cars. It disrupts whatever river traffic there happens to be on the Rhône and makes life miserable for harbor pilots in the port of Marseilles. It is the scourge of farmers, for it not only dries out the soil in an area where the rain is seasonal and irregular, but often attacking the fruit orchards in the spring, it strips the trees of their blossoms and breaks their branches.[10]

In other parts of the world similar winds sweep down from the north, sluice through narrow mountain passes, and debouch with sudden violence on temperate southern lands. The harsh winds of the bora are felt dozens of times a year along the Dalmatian coast from Trieste to Dubrovnik.

Just north of Cape Horn, where the prevailing winds blow with unusual force almost all the time, a small, capricious, but very intense gale known as the williwaw can suddenly appear out of nowhere, whistling through a narrow gap in the steep mountains that guard the Straits of Magellan.

In 1898 Joshua Slocum anchored his battered thirty-six-foot sloop *Spray* in a small snug cove in the Straits. He had just completed the first portion of a lone voyage around the world, had weathered a fierce gale for four days in the wild seas known as the Milky Way off the coast of Tierra del Fuego, and had brought his little ship into what he believed to be a safe harbour. He had got out his spare canvas and was attempting to improvise a new mainsail to replace the one that had been ripped to ribbons in the storm.

> The day to all appearances promised fine weather and light winds, but appearances in Tierra del Fuego do not always count. While I was wondering why no trees grew on the slope abreast of the anchorage, half minded to lay by the sail-making and land with my gun for some

game and to inspect a white boulder on the beach, near the brook, a williwaw came down with such terrific force as to carry the *Spray*, with two anchors down, like a feather out of the cove and away into deep water. No wonder trees did not grow on the side of that hill! Great Boreas! a tree would need to be all roots to hold on against such a furious wind.[11]

Sometimes the presence of mountains causes hot dry winds instead of the bitterly cold blasts like the williwaws. Inhabitants of mountainous regions have long been aware of a striking phenomenon that can occur at all times of year but is particularly frequent in the late winter. On a typical February day the little villages in the Alps are bedded in a crisp deep covering of snow. The peaked shapes of the wooden houses and the dark cones of the pine trees are etched against a brilliant white landscape. The mountain-tops seem startlingly clear against a deep blue sky.

Then all at once a line of high dark clouds begins to form over the mountain peaks. The clouds are flat on their lower edges and have gently curved tops with clean, smooth edges – lenticular or lens-shaped, these clouds are often called. A faint breath of warm air is felt, bringing with it the sound of a distant rumble from the high mountains like an old man grumbling in his sleep. The warm air flows stronger; the peaks and the cloud banks turn a deep purple and seem to come closer as the sky around them becomes an even more intense blue. Hot air rolls in now with a steady roar. Torrents of water from the melted snows pour down the mountainsides and avalanches threaten the villages. In a few hours the 'snow eater', or the föhn, has raised the temperature from several degrees below zero to 40° or 50°F (4° or 10°C). Snow and ice disappear as though by magic. In some cases it is not actually melted but sublimated, changing directly from crystals to invisible water vapour. And in spite of the wind's gluttonous consumption of moisture, the air is so dry that it desiccates everything it touches, sucking the moisture out of the furniture, the walls and the balconies of the frame houses. This drying wind causes a severe fire hazard in Alpine villages, where almost everything is made of wood. The inhabitants know by experience that when a föhn wind descends on a town, all fires must be extinguished, even cooking fires. But in spite of these precautions, fires do occur and a number of Swiss towns have been burned to the ground as a result of a visitation from the föhn.[12]

For many centuries the Swiss believed that their hot dry winds were a form of sirocco, the searing wind off the Sahara. But around the middle of the nineteenth century a Viennese meteorologist advanced a new theory for the origin of this wind. According to this theory, which is now generally accepted, the warmth and dryness of the föhn wind is caused by its excursion over the mountains. In Switzerland, for example, a low-pressure frontal system approaching from the north-west draws in air from the southern slopes of the Alps, and as this air ascends a mountain it cools off, losing its moisture by condensation. When it reaches the summit it leaves its last gift of moisture in the lens-shaped clouds that fly from the mountain peak. Then a portion of the air is drawn down the northern flank of the mountain, and as it descends, it is warmed. Its temperature increases one degree Fahrenheit (·6°C) for every 167 feet of height lost. With rising temperature the relative humidity of the air decreases and it arrives in the valley as a hot and very dry wind.

Föhn-type winds occur in many mountainous regions of the world. The dryness is a hazard to vegetation, particularly in the summer months and in southern lands where the föhn does not simultaneously produce quantities of water from melting snow. 'Leaves crumble to dust at the touch; corn, wheat and other cereals look as if they had been scorched by a fire; apples are described as being baked while hanging on the trees.'[13] The Santa Ana, a dry mountain wind that occurs frequently in the Los Angeles area, has often been the forerunner of fearfully destructive fires causing millions of dollars' worth of damage.

However, in some parts of the world the föhn wind is a blessing eagerly awaited by farmers and ranchers. In north-western Canada and Montana, without the warmth of these winds the snow would pile up week after week and month after month, endangering the survival of the great herds of cattle that must find forage on the ranges throughout the winter. The föhn is called the chinook in these places. Sweeping down over the east slope of the Rockies, pounding in like a rescuing posse arriving at the eleventh hour, it has saved untold numbers of famished and half-frozen animals. The chinook can melt several feet of snow in twenty-four hours and raise subzero temperatures to 40° or 50°F. Brown hills spring forth where just a few hours before snowdrifts had lain deep over the land and, with the unexpected warmth and moisture, new grass begins to grow. Because of the frequent occurrence of the chinook – thirty or forty times a year – cattle can be grazed on ranges all winter as far north as Alberta. Calgary has a mean winter temperature fifteen degrees warmer than Winnipeg, which lies farther south and is at a lower altitude. The beneficial effect of the chinook is even felt at Edmonton, three hundred miles east of the Rocky Mountains.[14]

The distinctive lens-shaped clouds that usually signal the approach of the föhn winds remain stationary over the mountains even when the wind is blowing at nearly gale strength. Constantly forming and dissipating high above the mountain peak and downwind of it, these clouds are markers of a remarkable standing wave pattern in the atmosphere. This wave pattern is similar to the ripples which occur on the downstream side of a large rock submerged in a river. The position of the troughs and crests of the wave remain in approximately the same position even though the water is flowing through the wave at quite a rapid pace. In the same way air passes through the standing atmospheric wave which occurs slightly downwind of the mountain-top. The air that sweeps up a mountainside reaches a height of several hundred feet before it begins to be deflected downwards. Then it enters into a series of undulations. Under just the right circumstances the crests of these waves can rise thousands of feet above the mountain peak. The size, shape, and position of these lee waves are determined by many factors – the shape of the mountain, the velocity and direction of the wind. The most remarkable lee waves in the world occur in the Sierra Nevadas near Bishop, California, generated by a fortuitous combination of circumstances: strong prevailing winds which blow directly across the range, the steep slope of the mountain which rises 9,000 feet above the floor of Owens Valley, and its unusually smooth shape on the lee side. Pilots of gliders can ride these waves like giant roller coasters.[15]

In 1954 the World Gliding Championship flights were held in New Zealand's Southern Alps near Mount Cook. The clouds that day dramatically outlined the lee waves, which stretched as far as the eye could see. Philip Wills, the world champion at the time, soared for many miles in a swift leapfrog hop from the ascending slope of one wave to the next, gaining enough altitude in each ascent for the hop across the down current.

In December of that same year Wills made a spectacular trip up into the lee waves of Mount Cook. He was not trying to set any records that day. He was just out for a practice run in a third-hand pre-war sailplane. Not expecting to fly particularly high, he had eaten a heavy meal a few hours before and was dressed in summer clothes. But that day he found advantageous soaring conditions, riding the lift of the lee wave that took him higher than he had anticipated. When he reached 30,400 feet Wills was still reasonably comfortable because the sun shone so intensely through the plastic cover of his cockpit and, since he was equipped with oxygen, he would have gone even higher, but at that moment he heard a noise like a pistol shot. Looking up, he saw that the cockpit cover had cracked from the severe cold outside. By skilful piloting Wills managed to descend and land safely. He described himself afterwards as 'a very dusty middle-aged gentleman, in summer clothes, nursing an uncomfortably bloated and borborygmal stomach . . .'[16]

Men like Philip Wills who have ridden these airy roller coasters up to the top of the troposphere describe the wonderful smoothness of the air in the lee wave:

> The unmistakable creamy smooth lift showed that wave-lift was about. This was it: I swung smoothly round into wind and hung almost stationary in space climbing silkily upwards. Overhead was a flat and rather dirty sheet of cloud, stretching as far as I could see in all directions. I found myself rising with gently increasing speed towards it, until in a short time it quietly took me in. The lift increased further and quite suddenly I burst forth into a scene that might have been on another planet.[17]

Some turbulence may be encountered just before entering the wave itself, then the air currents become very uniform with none of the sudden updrafts and downdrafts which are common in most clouds. This uniformity is responsible for the very distinctive shape and the firm cloud edges that are not torn or distorted by vagrant winds. The straight bottom edge of the clouds occurs at the condensation level, where the air temperature causes cloud droplets to form from the water vapour. The curved upper surfaces mark the contours of the standing wave.

One particularly extraordinary form of these clouds is characteristic of

19. 'Une pile d'assiettes'.

the lee waves over the Maritime Alps in south-eastern France. It looks like a stack of pancakes or – as the French say – *une pile d'assiettes*, a pile of plates. This stack of very thin round clouds is formed when the air is arranged in layers of varying humidity. The layers containing the most moisture precipitate into clouds and the alternate layers are clear. The reason for this striated formation is not completely understood, although the suggestion has been made that it could be caused by a jet wind slicing through a standing lee wave. *Une pile d'assiettes* is a dramatic demonstration of the way different atmospheric conditions can exist very close to each other in space. These differences are not rapidly dissipated by mixing processes. The layers preserve their own individual characteristics, forming part of the variegated structure of the atmosphere.[18]

The remarkable smoothness and predictability of lee waves make them favourite playgrounds for sailplane pilots. But once in a long while the lee wave may play a frightening trick on those unwary pilots who have been taking its gentle nature for granted. Under certain conditions a horizontal rotating cylinder of air can form directly under the crest of a very strong lee wave. [See Plate 20.] Curious cloud formations near ground level and swirling walls of dust had suggested to some scientists that extreme turbulence might sometimes occur under the lee waves. But most meteorologists were sceptical and it was the glider pilots who demonstrated the fantastic violence of these rotor waves.

At 10.33 a.m. on 25 April 1955, Larry Edgar took off from Bishop, California, in a Pratt-Read 195, one of the strongest sailplanes ever built. For several hours he flew the Sierra Wave, reaching a maximum altitude of 39,400 feet. There, the outside temperature was −64°C. Dressed in heavy wave-flying gear, he was well protected from these chilling temperatures. After spending two hours and forty minutes above 30,000 feet, the only discomfort that he experienced was that his feet were a little cold.[19]

It was just after 2 p.m. when he radioed Bishop that he was going to come down. Over Bishop and slightly to the east Edgar saw the leading edge of a large roll cloud and noticed little tufts of cloud out ahead of the main rotor:

20. A 'rotor' in a line of wave clouds is created by strong winds 9,000 feet above the floor of Owens Valley on the eastern flank of the Sierra Nevada.

These were forming downwind of Coyote, south of Bishop, forming and expanding very rapidly. It was apparent they might become a solid row ahead of the main roll-cloud. They might also move downwind and merge with the roll-cloud.

I had previous experience with short periods of flight in the rotor and had no desire to get any more. To avoid any cloud flying at this level, I turned northwest and increased my speed to about sixty-five to penetrate into the wind . . .

The north end of the little forward roll-cloud appeared to be approaching the leading edge of the main rotor cloud. It looked as though I would clear the top of the little tuft ahead, but it was building up very rapidly. Upwind, the area was clear of all clouds. I would be in the clear once I had crossed the narrow low spot in the roll straight ahead. To my right and left the cloud was continuous and larger.

My flight path went into the very top of the little cloud as it seemed to swell up before me at the last moment. I looked at the needle and ball. Suddenly and instantaneously the needle went off center about two thirds of the way. I followed with correction, but before I reached the neutral position, the needle violently swung the other way. Shearing action was terrific. I was forced sideways in my seat, first to the left, then to the right. At the same time the shearing force threw me to the right, a fantastic positive G load shoved me down into the seat – my head went forward and my chin was pressed hard against my chest. I could feel my body crumple in the seat as I blacked out. Then a violent roll to the left with a loud explosion was instantaneously followed with a crushing negative G load.

I was unable to see after blacking out, but I was conscious and felt my head hit the canopy with the negative load. There was a lot of noise, and I was taking quite a beating at this time.

I had made no movements on the controls except the rudder. Thoughts run through your mind very quickly at such moments. As my head struck the canopy, it felt like I had moved upward quite a way, and I wondered if I had accidentally unhooked or loosened my seatbelt. I was too stunned to bail out.

Then, just as suddenly as all this violence had started, it became quiet, except for the sound of the wind whistling by. I felt that I had been thrown clear of the glider. There was no question of falling, but rather one of being suspended out in space.

Something was holding both my feet. I tried to move them, but it had a firm grip. I tried to look at my feet and see what was holding them, but I still couldn't see – everything was black.

There seemed to be no twisting, shaking or tumbling in the fall. However, I must admit I was very much confused. I was still trying to squirm and pull my feet free, but I couldn't do it. I fell a way and then decided it best to try to open the parachute anyway. I felt and fumbled across my chest until I found the ripcord. I yanked; the parachute opened immediately – what a wonderful feeling!

At the same instant both my feet were free. My boots went when the parachute slammed open, so I was in my socks. Everything was black. It was quiet. All of this violence had taken place in just a few seconds. Now, for the first time I could really keep up with what was going on. If I could only see! My helmet, gloves, and oxygen mask were all gone. My feet were cool even though I had on three pairs of socks. The slippers went off with the boots.

I was concerned about being carried up to higher levels. Just a few minutes before, the rate of climb in 195 ahead of the roll-cloud had been eighteen hundred feet a minute. Still dark – there was a hissing noise. I felt down my right leg to locate my bailout bottle. I thought perhaps I could stick the hose in my mouth since the oxygen mask was gone. However, the hose from the bottle was broken off and completely missing. As I was exploring the bailout bottle with my hand, vision in my right eye returned – it was blurred but so helpful.

The first thing I saw was a faint light moving very slowly back and forth. It took me a moment to figure out this was the sun. I was in cloud, but it was not dense enough to completely blank out the sun.

I looked up at the parachute. It was a colorful thing with orange and white panels. There were some broken shroud lines. I looked down and noticed the ground through a little hole in the cloud. Now I realized what made the sun appear to move back and forth. I was turning and swinging on the parachute quite violently at times. Now and then the parachute would suddenly yank me upward.

I came out just below the main roll-cloud. It was a massive, dark, boiling thing. I didn't want to be carried upward so I pulled on the shrouds on one side to partially collapse the parachute. For the first time, free of the cloud, I could see parts of the Pratt-Read being carried upward past me. This was the first I had seen of the glider since hitting the turbulence, and it was my first indication that 195 had broken up in the air. I had not been thrown out because of a loose belt.

Seeing pieces of fabric and plywood going up and disappearing in the roll-cloud was quite an impressive sight, and I cannot express my feelings as I swung there on the parachute and realized these were pieces of 195. It may sound funny to some and inadequate to others, but at that time I exclaimed aloud, 'Darn!'

My left shoulder, arm and hand were numb and quite useless in tugging on the shroud lines of the parachute. However, I did manage to get my left arm up to my chest so I could use it to grasp the shroud lines that I pulled down with my right hand. Pulling with my right hand, it seemed like pulling on a big spring – when one relaxed at all, the parachute would pull back up.

The shroud lines pulled my left wrist in front of the left eye, so my wristwatch was right in front of my right eye. I wasn't particularly interested in the time, but since that was the thing I saw, I noticed the watch said ten minutes after three.

My right hand was bleeding profusely. Vision in my left eye was still gone, and I was concerned with having lost it as the left side of my face was all wet.

The wind was carrying me eastward over the valley. I was three or four miles south of Bishop. It looked as though I might land on the White Mountains to the east. I still couldn't tell that I was doing very well with the problem of getting down. I kept on tugging at the shrouds which was very exhausting. Then, gradually, the roll-cloud began to look a little higher and I could tell by the crest of the mountains that I was coming down.

I looked at my watch. . . . Ten minutes had gone by since I first looked at it. My right arm was becoming very tired of holding the shroud lines, but now I could see with my left eye! It was not possible to focus it, but I could see. I found it better to keep it shut and try to look with just the right eye.

I heard the BT towplane engine, and I was hoping they would see me. Al Langenheim did see me, reported my position to Bishop and flew around me as I landed. In the turbulent area, the wind would sound strong and whistle, then it would be quiet and calm as I hung from the parachute.

Near the center of the valley, four or five thousand feet above ground, the oscillation stopped. I was down in reverse flow. Now the wind was drifting me westward and a little north, very smoothly and rapidly. I let go of the shroud lines, totally exhausted.

Approaching the ground, I attempted to prepare for landing. My left hand, arm and shoulder were still numb and useless. I released the safety hasps over the harness quick disconnects as I drifted westward over the highway. I opened my left eye to try to help judge distance as the ground approached, but I was not yet able to focus the eyes together.

I was drifting backward across the ground quite rapidly, perhaps twenty or twenty-five miles per hour. I felt I should try to turn around by grasping across the shroud lines but still I could not raise my left arm high enough to accomplish this. I did grasp the fittings to disconnect the harness from the parachute, but I failed to act fast enough. I was stunned by the landing and do not recall being dragged on my face through the gravel.[20]

Larry Edgar recovered completely from this experience. After all the pieces of his sailplane had been found, the accident was reconstructed and the forces necessary to break up this rugged glider were estimated. The nose had been pulled off right at the seats by what appeared to be tension failure of the steel tubing. It was calculated that this would have required a force of about 16 Gs. The left wing and the tail boom had been broken off at altitude. Various control cables were pulled completely apart. The force required to cause these ruptures must have exceeded ten thousand pounds.[21]

Edgar's flight, part of a meteorological research project, had fallen victim to wind storm conditions far more violent than those normally encountered in thermal and ridge soaring. Yet even extraordinary experiences like this do not deter the real devotees of this thrilling sport. To ride on silent wings, to see the dazzling panorama of mountain ranges and cloud banks stretching clear out to the semicircular horizon that marks the edge of space, is an exultation that only a few have known.

These men have 'felt the sky with their fingertips'. They have known its silky smoothness, its tortuous turbulence, its waves of gently rising and falling currents. They understand – as no other human beings do – the responsiveness of the volatile medium that laps the mountain-tops and valleys of the earth.

21. *Soaring on the wind.*

CHAPTER ELEVEN

THE GREEN FLASH AND OTHER LIGHT SHOWS

Awake! for Morning in the Bowl of Night
Has flung the Stone that puts the Stars to Flight:
And Lo! the Hunter of the East has caught
The Sultan's Turret in a Noose of Light.

– Omar Khayyám[1]

One day in 1666 when Isaac Newton was twenty-four years old, he amused himself by performing a new experiment in his 'elaboratory'. Newton probably would have characterized this activity as another example of the special kind of self-indulgence to which he was addicted all his life. 'I do not know what I may appear to the world,' he said of himself many years later, 'but, to myself, I seem to have been only like a boy playing on the seashore, and diverting myself in now and then finding a smoother pebble or a prettier shell than ordinary, whilst the great ocean of truth lay all undiscovered before me.'[2]

On this day, he tells us,

> I procured me a triangular glass prism to try therewith the celebrated phaenomena of colours. And in order thereto, having darkened my chamber, and made a small hole in my window-shuts, to let in a convenient quantity of the sun's light, I placed my prism at its entrance, that it might be thereby refracted to the opposite wall. It was at first a very pleasing divertisement, to view the vivid and intense colours produced thereby . . .

In the screen that received the spectrum Newton then made a hole small enough to allow just one of the colours to pass through onto another screen, and he found that the colour of this separated ray could not be altered by reflection or by passing it once more through the prism.

> When any one sort of rays hath been well parted from those of other kinds, it hath afterwards obstinately retained its colour, notwith-

standing my utmost endeavours to change it. I have refracted it with prisms and reflected it with bodies, which in daylight were of other colours; I have intercepted it with the coloured film of air, interceded two compressed plates of glass; transmitted it through coloured mediums, and through mediums irradiated with other sorts of rays, and diversely terminated it; and yet could never produce any new colour out of it. It would by contracting or dilating become more brisk, or faint, and by the loss of many rays, in some cases very obscure and dark; but I could never see it changed in specie.

But the most surprising, and wonderful composition was that of whiteness. There is no one sort of rays which alone can exhibit this. 'Tis ever compounded, and to its composition, are requisite all the aforesaid primary colours, mixed in a due proportion. I have often with admiration beheld that all the colours of the prism made to converge, and thereby to be again mixed, as they were in the light before it was incident upon the prism, reproduced light, entirely and perfectly white, and not at all sensibly differing from a direct light of the sun . . .

After summarizing his observations, Newton ends with this sentence:

But to determine more absolutely what light is, after what manner refracted, by what modes or actions it produceth in our minds the phantasms of colours, is not so easie; and I shall not mingle conjectures with certainties.[3]

Three centuries and hundreds of thousands of experiments later we still do not know absolutely what light is. We know a great deal about it, but no one simple model is adequate to describe it. Light has many properties of a wave motion. We can measure its wavelength (the distance from the crest of one wave to the crest of another) and its frequency (the number of crests passing a given point in a unit of time). It has properties that are both magnetic and electric; so it is known as an electromagnetic wave. But light also acts in some ways like a stream of very tiny particles. Light travels through space at tremendous speeds, taking just a little over eight minutes to cover the 93 million miles between the sun and the earth. As a general rule, it moves in straight lines, deviating from this course only under special conditions. One of these exceptions is demonstrated in the spreading of light by Newton's prism.

When light passes obliquely through a medium where its velocity is slowed down or speeded up, the path of the ray is bent or *refracted*. An analogy may be helpful in understanding the phenomenon of refraction. Imagine a line of four or five girls walking arm in arm across a hard-surfaced field and approaching at an oblique angle a field buried deep in snow. The girl at one end – say, on the left – encounters the snow first and is slowed down by it. The line of girls, still arm in arm, pivots to compensate for the drag on the left. The rotation of the line continues until they are all walking in the snow. The angle of this rotation depends upon the difference in the speed of their advance on the hard surface and on the snow. A large difference causes a large angular displacement. In the case of light the difference in the velocities in the two media increases with the frequency of the wave. The higher the frequency, the larger the angle through which the light is refracted.

As Newton observed, this phenomenon of bending through different angles causes the ray of white light that passes through a prism to be spread out into a fan of many colours – red, orange, yellow, green, blue, and violet – always appearing in exactly that sequence because their frequencies are related to each other in that order. Red light has the lowest frequency of visible light, yellow slightly higher, and so on.

As light passes obliquely into a denser medium, the rays are bent towards a line drawn perpendicular to the surface. When they pass into a less dense medium they are bent away from the perpendicular. If light falls straight down on a medium whose surfaces are parallel, such as a flat plate of glass, it passes through without refraction. The wave front is slowed equally at all points and there is no bending; it emerges at the other side unchanged. The bright spectrum of colours remains concealed beneath the single intense impression – white.

The discovery of the principle of separating the sun's rays by a prism led eventually to the invention of the spectroscope, a remarkably sensitive instrument which enables physicists to analyse the radiation of the sun and other stars. Radiating bodies emit characteristic frequencies which make patterns of dark and light bands in the spectrum. These yield an amazing amount of information about the chemical composition, the temperature, and even the velocity of the object emitting the light.

Actually, the portion of the spectrum which we can see on earth is a very small segment of the electromagnetic radiation poured out by the

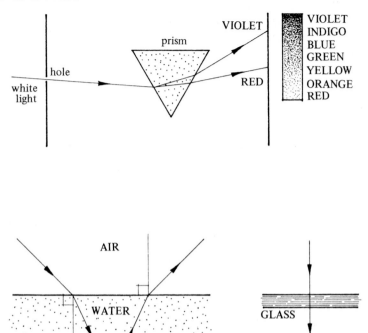

Fig. 8. *Three diagrams showing the behaviour of light:*
   (1) The separation of light rays – Newton's prism.
   (2) Light passing obliquely into and from a dense medium.
   (3) Light falling straight down onto a medium with parallel surfaces.

sun and the other stars. On the lower-frequency side, the infra-red, heat and radio waves cover a range of frequencies many times as wide as the spectrum of visible light. Beyond the violet are ultra-violet rays, X rays, and the extremely high-frequency gamma rays. Some of these rays are absorbed by our atmosphere. Virtually no X rays or gamma rays from sunlight reach the earth's surface and very little ultra-violet. The visible rays are partially absorbed, especially yellow and orange.

On the other hand, the higher-frequency visible waves, such as blue and violet, are strongly scattered by the molecules of the atmosphere, so blue radiation appears to come from everywhere, filling the whole day-time sky with light and obscuring the stars. As the balloonists and astronauts discovered, at high altitudes this softly diffused blue light of day gradually changes to deeper tones and the stars begin to shine through with a clear, steady light. (The twinkling we usually see is due to the turbulence of the air; so at great heights the stars do not twinkle, but the lights on earth do.) Finally, in the blackness of space sunshine is a sharp, blinding shaft of light.

The most energetic rays which reach the earth's surface are those to which our eyes respond and which we call light. If we could see infra-red or ultra-violet, the world would present a totally different aspect. It might look as it does to a honeybee or a luna moth. All mammals except man and a few monkeys and apes are colour-blind; everything appears to them in black and white or shades of grey. Bees, moths, and butterflies do not see red, but they do see ultra-violet. Birds, on the other hand, have a colour perception similar to ours. They can experience the whole range of tones from red to violet, witness the slowly unfolding colour show of each sunrise and sunset, and watch rainbows come and go across their flight path as they soar high in misty air.[4]

Is there anything in heaven and earth more magical than a rainbow? Appearing suddenly – out of nowhere – it flashes across a sky darkened by clouds and rain. Its arc of pure transparent colour, unattached to any kind of matter, seems ethereal, almost otherworldly. It is no wonder that the rainbow has been associated with divinity in every continent on earth, in myths and legends that go all the way back to the beginning of human experience. Many peoples thought of the rainbow as a bridge between heaven and earth. North American Indians called it 'the pathway of souls'; the Japanese named it the Floating Bridge of Heaven. Primitive Peruvians were so awestruck when a rainbow appeared that they remained silent until the magical sign had faded again from the sky.

The ancient Greeks deified the rainbow as Iris, the goddess who helped mankind by drawing the water up from the oceans and lakes and allowing it to fall again from the clouds as life-giving rain. She was also the swift messenger of the gods with golden wings and robes of many hues. Homer in the *Iliad* told the story of Aphrodite, who, fleeing wounded from a battle with the warrior Diomedes, was carried as fast as the wind along the arc of the rainbow to heaven.[5]

In Judeo-Christian faith the rainbow is the sign of God's promise to protect mankind from a recurrence of the Great Flood:

> And God said, This is the token of the covenant which I make between me and you and every living creature that is with you, for perpetual generations: I do set my bow in the cloud, and it shall be for a token of a covenant between me and the earth. And it shall come to pass, when I bring a cloud over the earth, that the bow shall be seen in the cloud, and I will remember my covenant, which is between me and you and every living creature of all flesh; and the waters shall no more become a flood to destroy all flesh.[6]

The writers of the Talmud, reluctant to say that the laws of nature had been suddenly changed at the end of the Great Flood, declared that the rainbow was one of the half dozen things created at twilight on the last day of creation.

The French mathematician and philosopher René Descartes was one of the first men to explain the rainbow. He correctly surmised that the presence of drops of water in the air caused a refraction of the sunlight that fell on them. To test this theory he made a large model of a raindrop out of a glass globe filled with water. Then he beamed a ray of light on the globe and was able to trace out the path of the light. As it entered the glass globe the light was refracted and, just as in a prism, the colours of the spectrum fanned out. When they struck the back side of the globe they were reflected back through the water again. The rays, doubly refracted, emerged from the front of the globe separated into the spectral colours and displaced about $42°$ (for red light) to $40°$ (for violet light) from the direction of the incident beam.[7]

In the conditions that create a rainbow the observer is between the sun and a shower of raindrops. If an imaginary line is drawn from the sun through the observer's eyes and extended straight on until it strikes the ground in front of him, all of the raindrops which are at an angle of $40°$ to $42°$ from this line reflect light back from the sun to the observer and, therefore, the rainbow describes an arc around this axis. The bow would be a perfect circle if it were not cut off by the earth. When the sun is very low, the arc of a rainbow is more than half of a circle, arching far up into the

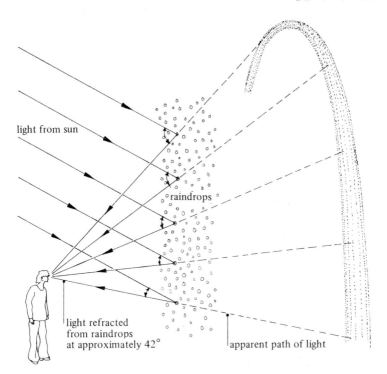

Fig. 9. Sketch showing the position of an arc of a rainbow in relation to the observer; different colours are refracted at slightly different angles: hence the rainbow's spectrum.

sky. As the sun rises higher a rainbow must subtend a smaller portion of the circle, lying nearer the horizon. At times when the sun is nearly overhead, it is not possible to see a rainbow at all unless the observer is elevated in relation to the horizon – for instance, on a mountain-top or in an aeroplane. Before sunset or just after sunrise, when a plane is flying between the sun and a rain cloud, a rare and beautiful sight may be seen by the passenger in the plane. The entire circumference of the rainbow circle is visible, with the shadow of the plane in the centre of the circle.[8]

Above the primary bow a second fainter bow is often visible a little higher in the sky. The sequence of colours is reversed, with the violet on top, the red on the bottom. This bow is created by two reflections within the raindrop. It is not as vivid as the primary arc because only a small portion of the light undergoes the second reflection.

Since the rainbow is an optical phenomenon, depending entirely on the angle of the light with the person who is viewing it, we can never find the pot of gold at the end of the rainbow. As we move towards it, it moves away. If we turn our backs on it and walk away, it follows us like a shadow. Elusive and ephemeral, the rainbow is the purest form of the many light and colour displays created by the refraction of light in the earth's atmosphere. Although the full range of colours is not always visible, the ones that do appear follow each other as predictably and logically as the notes on a musical scale. And the shape of the bow is a segment of a circle, the Greek ideal of perfection.

Sunsets and sunrises, on the other hand, can take an almost infinite variety of forms because of the many different cloud formations that add their own individual refraction patterns to the prismatic display produced when the sun's rays approach the horizon. As the light traverses a longer and longer portion of the earth's atmosphere, the rays are refracted into a spectacular panorama of colour. The red rays are bent and scattered the least and, therefore, the body of the sun itself looks red. Orange and yellow tones appear above the sun, somewhat weakened because they are strongly absorbed by the earth's atmosphere. Blue and violet are refracted highest into the sky and are scattered by the molecules of the air.

The glowing colours of a sunset are enhanced by the presence of clouds and veils of dust in the atmosphere. Clouds just below the horizon cause long shafts of golden light to shine high into the sky above the setting sun like a saint's halo. Clouds above the horizon are edged with fire. As the sun dips beyond the horizon a bright rose-red spot appears; it quickly grows larger and more diffuse, spreading upwards in translucent violet, pink, and salmon tones. The colours rapidly become more intense, filling a whole segment of the western sky and flooding the earth with a warm purple glow.

The intensity of this violet light varies from day to day. It depends upon the presence of very thin films of dust floating in the upper atmosphere. This layer is illuminated by the sun when it is already below the horizon, and the colour of the lowest portion of this beam is red since these rays have travelled through the densest layers of the air.

Following major volcanic eruptions, sunsets and sunrises are most spectacular, and the effects continue for several years. In the course of a few months after the eruption, these clouds of volcanic dust spread around a large portion of the planet.[9] The eruption of El Fuego in Guatemala caused remarkably brilliant sunsets which were first observed in southern Arizona but were later seen in many locations from Hawaii to France. The western sky was extravagantly coloured as the sun sank below the horizon and brightened again with a deep violet glow half an hour later.[10]

Sunsets and twilights seen from space are somewhat different from those viewed on the surface of the earth. Atmospheric structure and gradations can be seen much better than on earth and the colours are stronger.

During the manned orbital flights of the Russian Soyuz series, the earth's nocturnal, dawn, daytime, and sunset horizons were studied, with special attention given to the atmosphere in the transition zones where night is just turning into day or day into night. The astronauts' observations were described by two Russian meteorologists, K. Ya. Kondratyev and O. I. Smokty:

One of the most impressive sights seen from space is the earth's atmosphere near the dawn horizon. When a spacecraft is in the shadow zone and approaches the terminator line dividing the dark from the lighted regions of the earth, a crescent-shaped area colored differently at different heights is observed in the direction of the dawn horizon. Although the astronauts' description of the phenomenon differed from flight to flight, all were impressed by the variety and quick change of colors at space dawn.

According to visual observations made from *Soyuz IV* and *Soyuz V*, the earth's edge under these conditions is seen as a distinct black line. Above that the dawn aureole is red orange, which gradually turns into yellow orange, then yellow, then a narrow strip of dark blue of low brightness at a height approximately one-third the vertical size of the aureole. The dark blue strip is followed by an area of medium blue tones, which blends into light blue tones with a whitish tint. That area, occupying approximately two-thirds of the aureole height, changes in turn into black violet tones, which merge into the blackness of open space. The angular height of the dawn aureole is at its maximum immediately before the sun rises above the horizon. When the first solar rays appear, the aureole's height shrinks abruptly by approximately one-third, its brightness intensifies, and the dark blue strip disappears. Simultaneously, the saturation of the color tones of the dawn aureole increases.

If the sky is overcast, the lower part of the dawn aureole is seen as purple red instead of red orange. The spectrum of colors in the aureole is otherwise the same as in a cloudless atmosphere, but the upper edge of the cloud layer appears washed out and slightly phosphorescent. Rifts in the cloud layer are deep dark red tones.[11]

It is interesting that the Russian astronauts described space dawn as an arc above the horizon in the usual sequence of spectral colours, except that the blue band was wider and the colour green was missing. Green seems to be the individualist among the sun's spectral hues, sometimes refusing to take its allotted place in the usual parade of colours or, at other times, striking out as a loner and putting on a spectacular show of its own – as in the sunset and sunrise phenomenon of the green flash. Although this display occurs twice every twenty-four hours, it is so fleeting that it is witnessed only rarely on earth. Immediately after the sun disappears over the horizon in the evening (or before it appears at dawn), a vivid emerald green ray shoots up like a flame out of the horizon. This dramatic display fades so quickly that the phenomenon has often been dismissed as a fantasy or an optical illusion. In 1882 Jules Verne published a science fiction story, *Le Rayon Vert*, about a search for this mysterious ray and for many years it remained more in the realm of science fiction than of fact.

However, with the advent of colour film the objective existence of the green flash has been proven by photography. The display usually lasts for just a second or two and subtends a very narrow angle; so it is extremely difficult to capture on film. By the time the observer reacts and clicks the shutter, the ray is gone. Some successful amateur photographs have been made of the green flash by taking shots timed at exactly that second when the ray should appear and before it has actually been seen. The phenomenon has also been captured by professional scientists using a telescope and very fast film. D. J. K. O'Connell, S.M., director of the

Vatican Observatory at Castel Gandolfo, was able to obtain pictures showing the many diverse forms the green flash can take. The top rim of the disappearing sun is often coloured green, but the flashing ray itself is seen less frequently. Although a clear vital emerald green is the most usual colour, the ray sometimes shades into blue and even blue-violet.[12]

The isolation of the green colour results from the scattering and absorption of light in the earth's atmosphere. When the low red rays from the sun sink below the observer's horizon, they are blocked from view. Orange and yellow rays are largely absorbed by the atmosphere; the blue and violet are scattered by molecules of the air; green is the only strong colour left.

To see the green flash one must have an unobstructed view of the sunset. The horizon must be very sharp and the sky free of haze. A balloon floating in the stratosphere is an ideal platform from which to witness this unusual phenomenon, as Lieutenant Colonel David Simons discovered in Project Man High:

> It was 4:30 in the morning and the east was growing lighter. A reddish glow made a short fuzzy band where the sun would come over the horizon soon, and above it a layer of aquamarine faded away in the deep blue-purple of space. . . .
>
> Fervently, as an apostle awaiting the Lord's return, I watched for the sun to peek over the horizon. Soon it announced its coming, more beautifully even than Gabriel's horn could do.
>
> As I looked where the light of the sky was brightest, my eyes were caught by a brilliant green flash, a rarely seen phenomenon caused by the fact that blue-green rays of light bend slightly more than red rays and are therefore briefly visible slightly ahead of the sun's red rays as they bend across the curving earth. It was beautiful.
>
> The flash lasted less than a second before it was overtaken by the visible red rim of the sun edging over the horizon.[13]

On the earth's surface it is much more difficult to find favourable conditions for viewing the green flash. Clear air and a clean horizon are most often found on deserts, over water, or high on mountainsides. The duration of the flash depends upon the rate of descent of the sun below the horizon (or ascent at dawn) and this rate varies with latitude and season of the year. At the equator the sun sinks very fast and the flash lasts for less than a second. At the northern tip of Scandinavia, in the land of the midnight sun, the green flash lasts as long as fourteen minutes in mid-summer, when the slow sunset is followed immediately by a slow sunrise. The longest flash ever recorded was seen during Admiral Byrd's expedition to Little America in 1929. As the sun hovered near the jagged horizon of barrier ice, the green flash was seen intermittently for thirty-five minutes.[14]

At lower latitudes the observations of the green flash may be aided by certain unusual atmospheric conditions. Abrupt variations in the density and temperature of different layers in the atmosphere may act to increase the intensity and duration of the flash effect. A British astronomer, J. Evershed, reported seeing an emerald flash every evening at sunset on a sea voyage from Australia to Java. On each of these occasions there was also a mirage. Two images of the sun could be seen. As it set, the images approached each other and as they came together the green flash shot up into the sky. Evershed's explanation of this remarkable phenomenon was the presence of a narrow temperature inversion. Cooler, denser air in contact with the water caused the rays of light from the sun to be refracted as they travelled towards the observer. So an image of the sun, displaced upwards, was visible above the actual image of the setting sun. The inversion layer served to prolong visually the rate of the sun's descent below the horizon. This type of mirage has the effect of extending our line of vision over the edge of the world.[15]

All mirages are caused by the bending of light as it passes through regions of denser or more rarefied atmosphere. The path of the light is bent as it travels from the object to the observer as though it were passing through a large lens, but our minds are programmed to interpret visual impressions on the assumption that the light has travelled in a straight line. If there is a sharp increase in temperature with height (as in an inversion), a distant object appears to be displaced upwards. It may seem to be floating in the air and is often magnified. When the temperature decreases abruptly with height, the distant object appears to be displaced downwards in relation to the earth and the lowest portion of the image may be lost. A man, for example, looks as though his feet had disappeared below the curve of the earth.

Sailors and natives of high-latitude lands are accustomed to a very common type of arctic mirage which makes the ocean look flat or even

22. Mirage causes the miraculous effect of walking on water. The two boys on the left were actually walking on a sandspit. Both the sand and the boys' feet have disappeared below the optical horizon. The figures are also magnified.

concave upwards towards the horizon. Some scientists have surmised that this kind of mirage may have aided ancient explorers to find lands that lie beyond the true horizon, such as the Celtic adventurers who set forth from the Faroe Islands in their fragile skin-covered curraghs and sailed 290 miles to Iceland. In 874, after the Norse took possession of the Faroes, they probably followed the same route in their longboats. According to the old Norse sagas, Eric the Red heard a persistent legend in Iceland that more land lay beyond the sunset. Inspired by this story, he sailed west and discovered Greenland about 180 miles away. In these cases the arctic mirage may have aided the explorers, enabling them to see over the curve of the ocean surface and find the stepping-stones of land across the North Atlantic.[16]

But on other occasions the arctic mirages have played cruel tricks, creating wonderful landscapes out of thin air. A most striking example of this type of mirage deluded men for almost a hundred years and caused the launching of an expensive expedition to explore a land that did not exist.

In 1818 the British explorers James and John Ross set out to look for the North-west Passage to China. They sailed north of Baffin Island in Canada and, continuing farther west, they hoped to find a passage to the Pacific, but suddenly one morning they found the whole horizon ahead of them blocked by a great chain of mountains. There appeared to be no waterway through or around this formidable barrier. The Rosses abandoned the search and reported that there was no North-west Passage to China.

In 1906 Admiral Robert Peary, on one of his attempts to reach the North Pole, saw the same impressive chain of mountains rising above the arctic ice at what appeared to be about 120 miles west of Cape Thomas Hubbard. He wrote later that he longed to explore this spectacular land, which he christened 'Crocker Land'. But he did not allow this tempting prospect to divert him from his main objective of reaching the North Pole.[17]

Donald B. MacMillan was destined to have the frustrating experience of leading the expedition to explore Crocker Land. Financed by the American Museum of Natural History, the party set forth in the summer of 1913 and established a base at Etah on the coast of Greenland. But it was not until spring of the following year that the explorers, using sledges and dog teams, were able to reach Cape Thomas Hubbard, where Peary had sighted Crocker Land. Blowing snow carried on strong winds obscured the view, and without waiting for a glimpse of the land they had come to explore, Donald MacMillan and Fitzhugh Green set forth with several Eskimo guides and four sledges out over the treacherous ice of the polar sea. Even with air temperatures as low as $-25°F$ $(-32°C)$ there were leads of open water at midday interrupting the rolling hills of blue ice and in many places the frozen surface was dangerously thin, stretching and buckling like rubber. They had been travelling due north-west for several days when MacMillan noted in his diary: 'This morning Green yelled in through our igloo door that Crocker Land was in sight. We all rushed out and up to the top of a berg. Sure enough! There it was as plain as day – hills, valleys, and ice cap, a tremendous land extending through 150 degrees of the horizon.'[18]

The party pushed ahead now as rapidly as possible, fearing the imminent break-up of the ice. Once or twice, between snow showers and fog, they caught a glimpse of the mountain range glittering in the sunshine, but at other times they saw nothing but the endless expanses of ice. Finally on the seventh day they estimated that they had travelled 137 miles over the polar sea and were far beyond the place where Crocker Land was believed to lie. MacMillan climbed onto the summit of a pressure ridge. 'It was a perfect day,' he reported, 'the sky was a deep cloudless blue and all the mist had disappeared from the surface of the ice.' He swept the whole horizon with a powerful pair of binoculars – not the faintest semblance of land was to be seen. Reluctantly he was forced to the disappointing conclusion that what had so strongly resembled land had been simply a loom of sea ice. Crocker Land was nothing but an illusion.

The type of mirage produced by a layer of hot air next to the earth's surface is frequently seen in the desert. As the sand heats up during the day the air next to it becomes so rarefied that the light striking this layer obliquely is bent sharply upwards, almost as though it had been reflected in a mirror. Palm trees near the visual horizon and portions of the sky above them may appear as though they were being reflected in a large pool of water. At the same time, the normal image is seen in the usual way by the light that passes through a higher and less heated layer of air. As the observer moves forwards or backwards, so does the 'pool of water', which always remains just the same distance away from him. When the sun sinks low in the sky and the desert cools, the 'water' recedes towards the horizon and disappears. 'Bahr el Shaitan' (Lakes of Satan) these tantalizing mirages are called by the Arabs.[19]

A similar mirage is familiar to automobile drivers on hot summer days. Puddles appear on the road far ahead and disappear as the car approaches. These puddles are nothing more than pieces of the sky 'reflected' in the hot layer of air above the macadam surface.

Mirages can take many forms. They often cause wavy distortions like the reflections seen in the mirrors of an amusement park. They can literally turn the world upside down and give rise to tales that sound like Ripley's 'Believe It or Not'.

A story is told that one day in Cuxhaven, Germany, several children playing outdoors were alarmed to see an island hanging upside down over their heads. 'Mother, Mother,' the children screamed, 'an island is falling on us.' The sceptical woman glanced out the window and was

23. Halo around the sun.

amazed to see the familiar island of Heligoland with its characteristic red cliffs hanging over her head as though a gigantic hand had picked it up and inverted it in the sky.[20]

Citizens of Paris have reported several times seeing an upside-down image of the Eiffel Tower neatly balanced on the top of the real tower.

In the Straits of Messina when the sea is calm and the air is warm, a cloud can suddenly turn into a fantastic city with arches and towers and pillared porticos. The Italians named this type of mirage 'fata morgana' after the legendary Morgan le Fay, King Arthur's sister, who had magical powers of creating castles in the air.[21]

Clouds playing with beams of light can create many fairylike effects. When high thin cirrus forms an opalescent film across the sky, an enormous bright ring appears, making a perfect circle around the sun as though it were shining through ground glass. The ring is predominantly white with inner edges faintly red and outer edges blue.

These halos are caused by the diffraction of light by ice crystals acting like innumerable tiny prisms. The crystals must be very regular in shape and quite uniform in size to make such a beautifully symmetrical pattern. Since the majority of ice clouds are made up of a great diversity of snow stars and ice clusters, it is surprising that halos occur as often as they do – on the average about once every four days in the mid-latitudes. According to popular belief, a ring around the sun or moon foretells rainy weather. This prediction often proves to be correct, because cirrostratus clouds are frequently the forerunners of a low-pressure weather system.

Associated with the halos, concentrated spots of light sometimes appear at the same height above the horizon as the sun. These are called by many names: parhelia, sun dogs, or mock suns. Spectacular pillars of light and crosses centred on the sun also appear occasionally in the sky. The effects are all formed by reflection from certain alignments of symmetrical ice crystals and are most often seen at high latitudes.[22]

Coronae are similar to halos but much smaller. When thin fleecy clouds pass across the face of the moon, we can often observe coloured rings of light encircling it like an aureole. These rings of colour are formed

24. Two parhelia and the sun look like three suns in the sky.

by the diffraction of light by cloud droplets. Coronae can also be seen around the sun, but these are more difficult to observe because they are so close to the dazzling light.

While the sun and the moon may have their halos and their coronae, we human beings can have a glory around our own heads – or, more accurately, around the heads of our shadows. If you stand on a hillside when the sun is low in the sky and clouds are hanging close behind you, you can see your own shadow outlined against the mist. The head of the shadow is surrounded by a *glory* in vivid spectral colours. If you are standing with a friend, you can see his shadow, too, but the glory is only around *your* head. He, on the other hand, can see your shadow, but the aureole of coloured light encircles *his* head. This strange phenomenon is well known to the mountain climber, who calls it mountain spectre or Brocken bow, after the Brocken, a peak in the Harz Mountains where it is often reported. Another example of the glory is seen when a plane, flying towards the sun, emerges from a cloud. The passenger, looking back, can see the shadow of the plane 'caught in a noose of light'.[23]

In every latitude and elevation on earth a fascinating kaleidoscope of colour effects illuminates the sky. Dawn and dusk, the slow turning of the seasons – each brings its characteristic beauty as light from our star weaves its ever changing pattern of colour in the atmosphere. The polar regions, where light passes most obliquely through the many layers of earth's aura, experience the widest range of these optical wonders. The coming of a polar night to Antarctica was described by Admiral Byrd:

12 April
. . . It has been crystal clear, with a temperature of about 50° below zero, and a whispering southerly wind that sets fire to the skin. Each day more light drains from the sky. The storm-blue bulge of darkness pushing out from the South Pole is now nearly overhead at noon. The sun rose this morning at about 9:30 o'clock, but never really left the horizon. Huge and red and solemn, it rolled like a wheel along the Barrier edge for about two and a half hours, when the sunrise met the sunset at noon. For another two and a half hours it rolled along the horizon, gradually sinking past it until nothing was left but a blood-red incandescence. The whole effect was something like that witnessed during an eclipse. An unearthly twilight spread over the Barrier, lit by flames thrown up as from a vast pit, and the snow flamed with liquid color.

At home I am used to seeing the sun leap straight out of the east, cross the sky overhead, and set in a line perpendicular to the western horizon. Here the sun swings to a different law ...

The coming of the polar night is not the spectacular rush that some imagine it to be. The day is not abruptly walled off; the night does not drop suddenly. Rather the effect is a gradual accumulation, like that of an infinitely prolonged tide. Each day the darkness, which is the tide, washes in a little further and stays a little longer; each time the day, which is a beach, contracts a little more, until at last it is covered. The onlooker is not conscious of haste. On the contrary, he is sensible of something of incalculable importance being accomplished with timeless patience. The going of the day is a gradual process, modulated by the intervention of twilight. You look up, and it is gone. But not completely. Long after the horizon has interposed itself, the sun continues to cast up a pale and dwindling imitation of the day. You can trace its progress by the glow thrown up as it makes its round just below the horizon.

These are the best times, the times when neglected senses expand to an exquisite sensitivity. You stand on the Barrier, and simply look and listen and feel. The morning may be compounded of an unfathomable, tantalizing fog in which you stumble over sastrugi you can't see, and detour past obstructions that don't exist, and take your bearings from tiny bamboo markers that loom as big as telephone poles and hang suspended in space. On such a day, I could swear that the instrument shelter was as big as an ocean liner. On one such day I saw the blank northeastern sky become filled with the most magnificent Barrier coast I have ever seen, true in every line and faced with cliffs several thousand feet tall. A mirage, of course. Yet, a man who had never seen such things would have taken oath that it was real. The afternoon may be so clear that you dare not make a sound, lest it fall in pieces. And on such a day I have seen the sky shatter like a broken goblet, and dissolve into iridescent tipsy fragments – ice crystals falling across the face of the sun. And once in the golden downpour a slender column of platinum leaped up from the horizon, clean through the sun's core; a second luminous shadow formed horizontally through the sun, making a perfect cross. Presently two miniature suns, green and yellow in color, flipped simultaneously to the ends of each arm.

These are parhelia, the most dramatic of all refraction phenomena; nothing is lovelier.[24]

But one does not have to be an explorer, an astronaut, or a balloonist to enjoy the intricate interplay of light and cloud, sun and shadow. Even in the most heavily populated areas on earth the moving panorama of colour in the sky is there for all to see.

One day last October I took a late-afternoon flight from Chicago to Arizona. As the plane took off and swung in a northerly direction following the flight pattern over the airport, I saw against the white clouds on the right side of the plane a perfect circle of vivid rainbow colours and in the centre moved the shadow of our plane like a dark moth. The glory was visible for only a second or two as the plane turned in a great arc towards the south-west and flew into the sunset. Then began a long and brilliant display which lasted for almost an hour.

The western horizon looked perfectly clear and I watched for the green flash as the sun slowly sank below the horizon. I was rewarded with the sight of a little plume of green like a puff of smoke rising from the top of the sun's disc as it disappeared. The colour and shape of the flash were somewhat diffused, showing the presence of veils of dust or clouds which were not apparent to the naked eye.

Now the whole western horizon began to glow as though a brush fire were burning just below the rim of the earth. The flame radiated upwards and clouds of colour boiled up, laced with bright points of light like sparks thrown up from the hidden fire. A little higher in the sky, light vibrant red merged into orange, then golden yellow like the body of the sun itself. The coloured clouds rose higher, diffusing into paler yellow, their ragged edges blending into intense aquamarine, then deep dark blue where the main vault of the sky rode high and serene above the reach of the fire. Below us the land was dark as a mouse's skin; the little lakes and the winding rivers made puddles and ribbons of silver, mirroring the pale twilight.

As the display began to fade I glanced around me to see who else had been enjoying this wonderful spectacle. The woman at my left was deeply engrossed in a crossword puzzle. In the seat ahead of me a man had pulled the shade down over his window and gone to sleep.

CHAPTER TWELVE

WIND FROM THE SUN

Or the tiger sun will leap upon you and destroy you
With one lick of his vermilion tongue.

– Amy Lowell[1]

It was 7.21 p.m. Greenwich Mean Time on 4 October 1957, in Baykonur, Kazakhstan – a small bleak outpost on the grassy steppes of Russia north-east of the Aral Sea. This had been the heartland of Genghis Khan, whose horsemen had swept unopposed across these flat arid plains where only the bearded heads of grasses and a few poplar trees raised dark pointed fingers against the sky. Now in 1957 a strange monumental shape dominated the scene like a gigantic firecracker standing lonely and erect; it dwarfed the complex of low white buildings and concrete blockhouses grouped at its base. The team of engineers had just completed their final check of the systems. Sergei Pavlovik Korolev, the man in charge of the project, stood now in one of the blockhouses, his stock solitary figure framed in a window commanding a view of the launching pad. The countdown was proceeding inexorably to its finale: *tri . . . dva . . . odin* – and then an awesome burst of fire and smoke enveloped the base of the rocket. The lattice sway braces swung away like the petals of a giant flower opening and the pistil-shaped projectile lifted slowly, majestically into the sky. In the surrounding countryside the sound wave struck back with earthshaking fury and flocks of frightened birds rose from the brown grasses, filling the skies with confused flight as though they sensed that the whole character of the world in which they lived would be changed by that blast. Far above the birds, the rocket bored steadily higher; the first stage dropped away and Sputnik (the 'fellow traveller') arched into an elliptical orbit that would take it 583 miles into space, travelling 37 million miles and circling the planet earth 1,400 times.[2] The *New York Times* remarked, 'The creature who descended from a tree or crawled out of a cave a few thousand years ago is now on the eve of incredible journeys.'[3]

Night after night wondering people around the world watched for the satellite as it moved in its swift predictable track across the constellations. This first man-made star, which astronomers christened 1957 Alpha, was an aluminium globe about the size of a beachball and weighing 184 pounds. Filled with gaseous nitrogen and hermetically sealed, it contained two radio transmitters, instruments to measure the earth's magnetic field and the sun's energy. As radio stations in many countries monitored the insistent *beep . . . beep . . . beep*, scientists recognized that the irregularity of the radio signals meant that the satellite was sending back technical information about the layer of earth's aura that had never been systematically monitored before that time.

Sputnik I had been circling the earth for just a month when a much larger satellite was launched by the Soviets. On 3 November Sputnik II, weighing six times as much as Sputnik I, was put into a higher orbit, which reached an altitude of 1,055 miles at its maximum point. It contained a more complete array of scientific devices for measuring the solar radiation in the shortwave, ultra-violet, and X-ray regions of the spectrum, as well as cosmic radiation, temperature, and pressure. Sputnik II also carried a passenger – Laika, an eleven-pound black-and-white female dog of uncertain breed. Laika was truly the first astronaut and, like all those to follow her, she had undergone a period of careful preparatory training under satellite conditions. When she took off on her historic journey, she was housed in an airtight compartment that was equipped with an air-conditioning system. Oxygen was provided from a large tank and the air was regularly cleansed of exhaled carbon dioxide and water vapour. Laika was artificially fed with food suspended in gelatin. Electrodes and sensors monitored her vital processes. A photograph published by the Soviet press showed Laika strapped into her box and attached to her scientific life-support system. Her large dark eyes looked sad, apprehensive, infinitely patient; her long pointed ears drooped down at the tips. [4]

Like John Wise's kitten, Nellie, who 130 years earlier had embarked on a similar adventure, Laika was an unwilling participant in a scientific experiment designed to test whether an organism as complicated as a dog or a cat could survive in the strange environment so far above the earth's surface. But Laika was not as fortunate as Nellie, who limped home two days after her trip into the thunderstorm. No provision had been made for bringing Laika safely back to earth and there are conflicting stories about the nature and timing of her death. Soviet Premier Bulganin told reporters on 11 November that Laika had been alive on 10 November when Sputnik's radios suddenly stopped operating. At a press conference on 15 November Dr Alexii Pokrovsky, the physiologist who had helped train Laika, said that she had died 'an absolutely painless death' brought about by the failure of the oxygen equipment.[5] But with all communication severed on 10 November, who knows what really happened to Laika in those last days? Did the food supply give out before the oxygen failed? Or did the air-conditioning system cease to function first, allowing Laika to be slowly roasted in the intense heat that must have built up in the satellite during its daytime exposure to solar radiation? At night temperatures would have dropped low enough to freeze her solid within a few hours. In any event, Laika could not have survived long after any one of these systems failed. For five months thereafter the silent satellite carrying her corpse continued inexorably to lace its prescribed path across the night sky. On 14 April 1958, the Soviet news agency Tass announced that Sputnik II had disintegrated that morning. The pieces dropped into the dense atmosphere and were immediately consumed in in flames. The same fate had overtaken Sputnik I on 4 January.

Meantime, the spectacular success of the Soviet space scientists spurred feverish activity in the United States, where the Navy and the Army had been working on competitive rocket systems for long-range ballistic missiles. There was a plan (announced in 1955) to put a satellite into orbit during the International Geophysical Year, 1957 to 1958. This plan, however, had been delayed because of political manoeuvring between the two military services and other branches of government. At the time of the launch of Sputnik I and II, the Navy had been given the responsibility for placing an earth satellite into orbit during the IGY, using the Vanguard rocket. This plan was now speeded up and in the first week of December 1957 the attempt was made. It was a highly publicized event; for weeks newspapers, television, and radio had been pouring out a constant barrage of stories about this historic experiment. Several times the launch was delayed, and then at 1.00 a.m. on 6 December the final countdown began.[6] William R. Shelton, a reporter who was an eyewitness on this occasion, described the scene:

Through binoculars we saw a rime of frost building up on Vanguard's first stage as the subzero L O X (liquid oxygen) was pumped aboard. At 11.44 a.m. the rocket's umbilical, or service, cords dropped away. Launch was imminent. . . .

Suddenly a splayed tongue of flame darted from the base of the rocket. Was this the fire of ignition – or something else? Vanguard began to rise. Then it stopped. It began to sink back. As it fell into its own launch ring, its thin skin ruptured. Instantly a mass of incandescent red flame and boiling black smoke engulfed the rocket. The exploding fireball blossomed 70 feet high as L O X and kerosene combined in a fiery demise.[7]

This disaster, witnessed around the world, was a terrible blow to the prestige of the United States. In the weeks that followed, the government decided to give highest priority and backing to a group of rocket engineers who had come from Germany at the end of the Second World War. This team of experts, headed by Wernher von Braun, were employed as civilians by the U.S. Army. They promised that they could put a satellite into orbit within ninety days. Actually they did much better.

On the night of 31 January a Jupiter C rocket stood on the brilliantly lighted launch pad at Cape Canaveral, its ice-encrusted form reflecting the criss-crossing beams of the searchlights. The countdown that night proceeded without a flaw. At 10.38 p.m. the rocket's first stage ignited. Then with perfectly controlled and deliberate power it rose on a cloud of flame, drew a brilliant arc of flight high in the night sky, and carried America's first satellite, Explorer 1, even farther than had been anticipated. It sailed into an orbit that reached 1,500 miles at its farthest point from the earth.[8]

This American satellite, weighing just 31 pounds, was much smaller than Sputnik I and II, but it was packed with a sophisticated array of scientific instruments designed under the supervision of James Van Allen and assembled by the staff of the Jet Propulsion Laboratory at the California Institute of Technology.[9]

Official jubilation and relief greeted the success of the launching. In spite of the Russian lead, America had not completely lost face. The state of U.S. technology, the skill of her engineers and scientists, even the nature of the free enterprise system, were vindicated by the shiny cylinder threading its way through space.

The true vindication, however, became apparent several weeks later when the radio signals that were pouring back to earth from the satellite had been analysed by James Van Allen and his team of scientists. Explorer I had been well named. It was providing dramatic new information about the nature of space far above the main body of the earth's atmosphere. In Van Allen's words, it was 'painting a broad, deep stripe of knowledge 4,700 miles wide and 1,400 miles thick around the waist of the world.'[10]

Up until the time when these satellites went into orbit the characteristics of the earth's environment above an altitude of approximately 120 miles had been only guessed at by measurements made on the ground. The difficulties of interpreting information obtained under these conditions can be compared to studying the troposphere from the bottom of the sea. Imagine that a species of intelligent aquatic creatures, living on the ocean floor, attempted to analyse the nature of the air above the water surface. Let us suppose that they had at their disposal sensitive devices that could send signals up to the water surface and into the air itself. They could measure the way these signals were reflected or scattered by the atmosphere they encountered. By means of these measurements our hypothetical species would be able to gain some understanding of the kind of matter that existed far above their habitat. But they would have little or no information on the distribution of that matter – no knowledge of the complex circulation of wind currents or their interaction with the waves on the ocean's face, no understanding of the cyclic changes of the seasons, no explanation of the turbulence brewed by violent storms hundreds of miles away. In much the same manner, human beings had probed with radio and radar beams the space far above the gaseous envelope of the earth. They had studied the light rays that penetrated through the atmosphere and had ingeniously applied their intelligence to construct a total picture from these scraps of information. It is no wonder that this picture was dramatically revised as soon as instruments were lifted into that medium itself and were operated continuously for long periods of time. A door was suddenly opened, revealing a much more complex and interrelated system than had previously been imagined.

Before the late 1950s it had been believed that the atmosphere gradually thinned out, merging into space, which was virtually devoid of matter. It was known that occasionally showers of charged particles, ejected from the sun, collided with the earth's upper atmosphere, causing magnetic storms and displays of aurora borealis, but these were considered to be isolated events rather than part of a regular and continuous process. It was assumed that the magnetic fields of the sun and the earth did not extend far enough to affect the gases surrounding the other body. In fact, the earth and the sun were seen as two separate bodies related only by gravity and radiation across the empty space between them. After Explorer 1 went into orbit this picture was greatly altered.

Now it is understood that the sun and the earth are part of a much more closely integrated system. The earth is continuously immersed in a stream of charged particles flowing outwards from the sun. This stream is called the *solar wind* and, like the winds on earth, it is variable, blowing sometimes very hard, sometimes softly.[11]

The solar wind consists mainly of ionized hydrogen – protons and electrons having positive and negative charges respectively. A gas that is broken down in this manner is known as a *plasma*. The plasma of the solar wind flows radially outwards from the sun with a velocity that ranges between 220 and 500 miles per second, taking on the average about three and a half days to cover the distance between the earth and the sun. The solar wind is almost incredibly tenuous and hot. It has an average density of about 90 ions per cubic inch. (In comparison there are approximately 443,000,000,000,000,000,000 molecules in every cubic inch of atmosphere at sea level.) The temperature of the ions is about 300,000°F (160,000°C). This extreme temperature, of course, does not mean that a spaceship or a satellite would be melted by the heat of the solar wind. Temperature is a measure of the average velocity of the particles and in the solar wind this velocity is very high. But since there are so few particles per cubic inch, the amount of energy transferred when the solar wind strikes a solid body is small.

Borne along with this wind of charged particles, the magnetic field of the sun is still strong enough when it reaches the earth's surface to be a significant factor. Just as a current of air or water sweeps around any solid object in its path, the solar wind envelops the earth. It distorts the geomagnetic field (called the *magnetosphere*), bending it into a long parabolic shape similar to the bow wave of a boat. In the direction away from the sun the magnetic field streams out under the pressure of the solar wind in an elongated tail which extends far beyond the orbit of the moon. Between the earth and the sun a shock wave is formed some 40,000 miles from the earth. Most of the solar wind does not penetrate beyond this point. The interaction of the magnetic field of the earth and the field carried by the solar wind sets up regions where charged particles oscillate back and forth, trapped in regions of space called the Van Allen Belts, after the scientist who was most prominently involved in their discovery.[12]

Instruments carried aboard Explorer 1 and later Explorer 3 revealed amazingly high counts of energetic charged particles in two bands that are centred over the equator at altitudes around 2,000 to 4,000 miles and 12,000 to 18,000 miles.

Within the charged layer the trapped particles move in tight cylindrical helices along magnetic lines of force as though they were prisoners condemned forever to run up and down spiral staircases, and the whole aggregate of these imprisoned ions drifts slowly around the planet. The electrons move from west to east and the protons move from east to west. Since an electric current is nothing more than a stream of moving electrons, there is, in effect, an electric current circling the planet.

Some of the imprisoned ions eventually work their way out of the magnetic trap. The weakest portion of it occurs along the night side of the planet at the place where the Van Allen Belts come nearest to the earth. The particles that escape in this way are dumped into the ionosphere, where they interact with the atmospheric ions to form complete atoms and liberate large amounts of energy.

Polewards of the weakest portions of the Van Allen Belts is a region where the magnetic field of the earth does not trap or deflect the particles of the solar wind at all. They spiral directly in towards the poles along the lines of magnetic force until they strike the ionosphere. Here they also interact energetically with the thin gases. These invasions of the ionosphere raise its temperature to very high levels. At altitudes of 63 miles temperatures of 1,600°F (870°C) have been measured over the equator and 2,600°F (1,427°C) over the North Pole. The energy pouring into the ionosphere also generates brilliant displays of aurora borealis or aurora australis, the 'northern and southern dawn'.[13]

The similarity to sunrise implied in these names is misleading; aurora displays usually begin a few hours after sunset and are all over by dawn. An observer on earth looking towards the dark northern horizon sees it begin to glow with a pale greenish light. It brightens rapidly; white rays faintly tinged with magenta and lime green arch high in the sky.

25. Photographs of Aurora Borealis.

More and more rays of light appear; they move, break up, and re-form in wavering luminous sheets like firelight flickering on the ceiling of the world.

Throughout recorded history tales have come down to us describing the awe and wonder inspired by these nocturnal displays. In far northern lands where aurora borealis is almost a nightly occurrence, it became part of the folklore and religion of the people. Some tribes believed that gods carrying flaming torches were duelling in the dark heavens. To the Eskimos in Hudson Bay, the polar lights are cast by lanterns carried by demons who are searching the universe for lost souls.[14]

But farther south, where such displays are very rare, they caused great fear. The story is told that in 1585 on a clear autumn night thousands of terrified French peasants left their homes in the peaceful countryside and crowded into the churches to pray. The sky was on fire! Curtains of flame rolled in over the northern horizon and beams of white and red light shot high as from a gigantic explosion just over the edge of the world. Shortly after midnight it seemed as though the prayers of the people had been answered. The firelight in the sky slowly subsided and soon the peaceful dome of heaven was once more studded with stars. Gratefully the people left the sanctuary of the churches and crept back to their cottages.[15]

Observers who have not been overcome by terror have enjoyed watching the rapidly changing shafts of light and appreciating the beauty of the auroral display. In 1916 a traveller described northern lights seen from a dog-sled as he was pulled across the snowy wastes of far north-east Siberia:

Twilight had gone, and night's profound darkness had fallen. Directly overhead sparkled the brilliant pole-star, the eye of the 'Shepherd of the Heavens,' as the natives say, and further away were the three deer with their Lamout hunters (Orion). Presently there appeared a faint whitish glimmer in the northeast, and in an instant, as though traced in phosphorus, an arch of light flashed across the black sky. Along the whole extent of the arch ran translucent waves of light, colourless and phantom-like as in a dream. Suddenly there spurted out a vast wavy river of multi-coloured fire, the arch glowed with a green brilliance, and far out over the entire expanse of the heavens rays of blood-red light were flung, reaching to the 'sieve of the beautiful Ourgal' (Pleiades). It seemed as though the heavens reflected the gigantic fires lit on the shores of Arctic Ocean by the giants Oulahan-kigi, who, according to native legends, have their dwelling place there. Suddenly the arch was extinguished, and only a few greenish points of light flashed here and there on the horizon. So wonderful and awe-inspiring is the sight of the Northern Light that one feels the truth of the native proverb: 'Who looks long on the heavenly light becomes mad.'[16]

The fact that auroral displays have caused concern and comment over the years has resulted in an extensive historical record of these events which is complete enough to help scientists identify the time and place of their occurrence over a span of several centuries. From these ancient records and continuous observation throughout the past hundred years

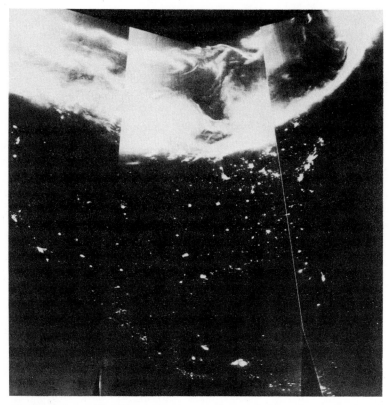

26. *Composite satellite photos taken in 1974 show oval shape of aurora circling the north geomagnetic pole. City lights of North America can also be seen.*

we know that auroral lights usually reach their maximum brilliance about midnight. They occur with the greatest frequency in a belt about 23° from the magnetic pole in each hemisphere. Between 60° and 45° they are seen frequently, but below 45° they are rare.[17]

One of the most significant discoveries about aurora lights is that their incidence waxes and wanes in eleven-year cycles. This fact provided one of the clues that suggested their origin. It had also been observed that certain mysterious dark spots on the sun increased and decreased on a cycle which varied between eight and fourteen years but averaged just slightly over eleven. When these two phenomena were plotted together, a remarkably close correlation was found, and these facts led scientists to conclude that aurora lights are caused by jets of charged particles emitted by the sun. When they strike the molecules of the upper atmosphere they initiate reactions which result in the emission of light.

Chinese observers using the unaided eye had observed sunspots almost a thousand years before they were definitely identified by scientists in the Western world. In 1611, after the telescope was invented, Galileo studied these strange markings on the sun and described them in a book that bore the title *History and Demonstrations Concerning Sunspots and Their Phenomena*. (Unfortunately, Galileo's studies of the sun damaged his eyes, which had already been impaired by infections in his youth. During the last four years of his life he was completely blind.)

However, the eleven-year cycle of sunspot activity was not noticed until the late eighteenth century and was actually confirmed by Heinrich Schwabe in 1843. As a matter of fact, the cycle itself did not occur in a regular manner during the hundred years after sunspots were discovered. Throughout that century the sun seems to have behaved very strangely. The old records of sunspots, which are very sparse and, therefore, not completely conclusive, show that two maximums may have occurred in the years immediately following Galileo's discovery in 1611, the peak times coming about fifteen years apart. Then the intensity of sunspot activity declined and remained very low until 1715. During that time polar lights were extremely rare even in such northern cities as Copenhagen and Stockholm and when they did occur they caused great consternation.[18]

Throughout these seventy years the paucity of auroral displays was only one of several indications of decreased solar activity. Normally, a

27. Streamers flaring out from the main body of the sun can be seen best during a total eclipse.

rim of flaming gases, with streamers extending thousands of miles above the solar disk like the petals of a flower, can be clearly seen during total eclipses of the sun. This is known as the solar *corona* (which should not be confused with the rings of coloured light that sometimes appear around the sun because of light diffraction by our own atmosphere). These encircling clouds of fire were not reported in total eclipses during the seventy-year period of the quiet sun.

Another interesting piece of evidence was turned up by A. E. Douglass during the course of his pioneer work on tree-ring dating. He had noticed a cyclical change in the size of tree rings. In general they tended to increase and then decrease again over a period of a decade or two. But when he looked for the same pattern in wood samples from the seventeenth century, he found little or no evidence of cyclical changes in growth. Later he read the reports of low sunspot activity during those years and he concluded that this must have been the reason for the unusual uniformity in the tree rings.

The relationship between sunspot activity and weather is still an open question and will be discussed in a later chapter. However, there is no doubt about the fact that auroral lights are directly related to bursts of solar wind which accompany increased solar activity and which, like the sunspots, seem to follow approximately an eleven-year cycle.

Since the first satellites went into orbit much has been learned about the solar wind and about the sun itself. A sunspot is now known to be a shallow depression in the surface of the sun. It may cover an area more than three hundred times the entire surface of the earth. It is dark in colour because it is cooler than the surrounding solar matter, but the temperature of even this 'cooler' area is 6,500°F (3,600°C), far hotter than anything we can imagine. The rest of the surface is an incredible 10,000°F (5,500°C) and some of the gases in the solar corona reach temperatures of 1,800,000°F (1,000,000°C). The area of comparatively cool solar surface in a sunspot is surrounded and confined by a very intense magnetic field. [19]

Throughout the eleven-year cycle not only do sunspots increase and decrease in number but the position where they appear changes in a regular manner. At first they usually occur in a band about 40° north and south latitude. As the cycle continues they make their appearance closer and closer to the sun's equator until finally they are occurring within 5° to 10° from it. When these spots fade out new ones appear again in the 40° band. Strangely enough, the new spots have magnetic polarity opposite to those of the previous cycle. In fact, the magnetic fields associated with the sun are extremely complex and they regularly reverse their polarity.

Most people think of a magnetic field as something fixed and invariant like the fields produced by the simple bar or horseshoe magnets that are familiar fixtures of high school science labs. To a casual observer the magnetism of the earth appears to have the same stable characteristics. The compass needle unfailingly points to the geographical North Pole and has been guiding explorers and navigators since two centuries before

Columbus discovered the New World. But a closer examination reveals that the earth's magnetic field is neither permanent nor invariant. There are many small fluctuations in it from time to time and from place to place. The earth's poles appear to have 'wandered'. At least they have moved in relation to the land masses and there is very good evidence that complete reversals have occurred, with North and South poles suddenly changing places. A reversal of this kind occurred 700,000 years ago, a time when primitive men were chipping simple stone tools and hunting animals on the savannas of Africa. Biologists do not have sufficient information at the present time to decide whether this reversal adversely affected the development of the human species. But sediments taken from deep layers of the glaciers in Antarctica have given indications that mass die-offs among species of micro-organisms occurred near the time of magnetic reversals, and these extinctions may have been caused by important climatic changes. These discoveries are far from conclusive; much further work in this field is needed to clarify the role of the earth's magnetic field in biological history. [20]

It is now believed that the magnetism of both the sun and the earth is caused by the motion of electric currents, just as magnetic fields are produced in a dynamo. Any moving stream of charged particles creates a magnetic field in the space around it, and a current flowing in a coil around an iron core, for example, can produce very powerful magnets whose field changes with variations in the strength and direction of the electric current. Within the nuclear holocaust of the sun every flow of matter carries great quantities of electric charge and produces electric and magnetic fields. These streams of solar plasma move in intricate patterns, much as our winds move on earth, but in a much more energetic and violent manner. Some of the extremely hot gas particles are moving at speeds great enough to overcome even the sun's very high gravitational field. Jets of these energetic ionized particles flow out into space, and as the sun turns on its axis each jet sweeps away from the sun in a curved stream like water spurting from a rotating garden hose. This stream intercepts the earth and passes on, returning again in twenty-seven days when the sun (as seen from the earth) has completed its rotation. Since there are many such streams emerging from the sun, the solar wind intercepted by the earth is a continuous but varying phenomenon.

Once in a long time an extremely intense flare of matter erupts from the surface of the sun, usually appearing at the locations of a rapidly changing sunspot group. The flare may be so bright that it is visible against the dazzling disk of the sun, with flaming filaments extending as far as 60,000 miles across the sun's face. The particles reaching the earth from such a solar flare are about one thousand times more energetic than the normal solar wind. An eruption like this causes the most spectacular aurora displays and a 'magnetic storm' which disturbs radio communications and compass needles on earth. A large flare may liberate more energy than a million H-bombs in the form of high-speed charged particles and powerful beams of radiation. Following a major solar flare soundings of the stratosphere at heights of about 100,000 feet have shown sudden increases in temperature as much as 70°F (39°C).[21] Episodes of this kind represent a grave hazard to astronauts. If such a flare occurred when they were travelling in space beyond the protective shield of the earth's atmosphere or when they were walking on the moon, they would be exposed to dangerous doses of radiation. Major solar flares occur a few times a year. Fortunately, up to this time, there has been no occurrence of such a flare during a manned flight.

Throughout the past decade much interesting information has been coming in from satellites that orbit the earth above the distorting influence of the atmosphere and take telescopic pictures of the sun in X-ray and ultra-violet radiation as well as visible light. One surprising discovery revealed by photographs from Skylab is that small eruptions, similar to flares but much less intense, occur every few hours on the solar surface. These frequent eruptions, which are visible only from a platform above the atmosphere, probably account for the small fluctuations in the intensity of the solar wind and the geomagnetic field.[22]

Experiments carried aboard the Apollo missions to the moon also provided information on the solar wind. The solar wind experiment which was set up by Neil Armstrong and E. E. Aldrin, Jr, on the lunar surface consisted of a piece of aluminium foil one foot wide and four and a half feet long positioned perpendicular to the solar rays. It was exposed for 77 minutes; the aluminium foil was then rolled up, returned to earth, and examined in the laboratory. On the Apollo 12 mission the experiment was repeated with an 18·75-hour exposure. As expected, particles from the solar wind had penetrated into the roll of foil. The size and shape of their tracks revealed their identity. Hydrogen ions constituted by far the largest proportion of these particles; about 4 per cent of the total were helium ions. A few neon and oxygen ions were also present. The distribution of the particles in the roll of foil showed that the rays were highly directional, coming straight from the sun to the lunar surface undeflected by any magnetic field. All of these facts confirmed the accepted theories about the solar wind.[23]

But none of the discoveries so far has thrown much light on the cause and development of sunspots. 'Their behaviour is still not understood in spite of a number of "explanations" that have appeared in the scientific literature,' says E. N. Parker, the scientist who is most responsible for the development of the theory of the solar wind. 'With further knowledge has come more mystery . . .'[24]

Here on earth all forms of life are totally dependent upon the sun for heat and energy, but without the complex screens and filters provided by the various layers of earth's aura, life on this small planet would not survive the fierce embrace of its star. Exposure to a series of solar flares could wipe out most of earth's living organisms and render the surviving ones incapable of reproduction. The aura acts like a translucent skin protecting the earth from the raw impact of the solar environment.

The magnetic field of the earth is the first screen that intercepts the flow of matter and energy from the sun. Most of the solar wind is deflected away from the earth at the edge of the magnetosphere. A small percentage of the incident particles enter the Van Allen Belts and are trapped there. Only at the unprotected poles do charged particles find their way easily into the earth's atmosphere, causing magnetic storms and the shimmering curtains of auroral lights.

The magnetosphere, however, does not protect the earth at all against the X rays and ultra-violet light that pour down upon it from the sun. Magnetism acts only upon charged particles, and radiation passes through it unhindered. The ionosphere is the first shield against these life-destroying forms of radiation. Those rays with the highest frequency expend their dangerous energy when they strike molecules of the tenuous upper atmosphere, transforming these molecules into ions and, incidentally, performing another benefit for mankind by creating a charged layer which reflects radio waves, making long-distance communication possible. Below the ionosphere there is another vitally important shield, the ozone layer (see Chapter 6), which absorbs most of the energy from the ultra-violet component of the radiation which has passed through the ionosphere. Finally, the main body of the atmosphere acts as the ultimate filter, reducing the small component of ultra-violet and charged particles that have managed to penetrate to this level, and letting through the beneficial, life-giving light and infra-red heat waves. Taken together these layers provide a remarkable shielding system. Any inadvertent weakening of this beneficial screen that protects life on earth must be viewed with the greatest concern.

In 1975, scientists at Stanford University were experimenting with radio emissions from such natural sources as lightning flashes, observing the manner in which these signals were channelled along the earth's magnetic field by electrons oscillating in the Van Allen Belts, and they discovered a curious phenomenon. The movement of these spiralling electrons was apparently being influenced by some radio-type signals generated on earth. The signals were identified as coming from electric transmission lines in Canada and the eastern United States, where a large grid of high-voltage lines seems to be acting like a giant antenna sending low-frequency signals into the ionosphere. The energy in those impulses is amplified a thousandfold by the energetic particles in the Van Allen Belts, causing some of the particles imprisoned there to be dumped into the ionosphere.[25]

When this discovery was reported in the press, Robert A. Helliwell, who headed the research, was quoted as saying that this discovery opened the way for ground-level manipulation of the radiation belts and the radio-reflecting layers of the upper atmosphere. Taking a long leap from this statement, some reporters declared that the earth's radiation belts were now 'under human control'.[26] This statement is far from the truth. Radiation belts are being inadvertently altered by transmission lines and possibly by other activities of man. It is a long and difficult road between the discovery of such inadvertent effects and true control in the sense of understanding the total process and altering it in a way that is known to be beneficial. Inadvertent changes of the earth's atmosphere are so common that they hardly make news anymore. Real control of our environment is rare indeed.

The discovery of the solar wind was one of the major scientific break-throughs of the space age. We know now that matter from the sun extends far beyond what we had previously imagined. Our planet moves, lives, and breathes in this charged medium. Earth's aura does not come to an end with the last tenuous gases of its atmosphere; it extends far out beyond the ionosphere and even far beyond the orbit of the moon. In fact, no dividing line can be drawn between the aura of the sun and that of the earth – they flow together, bending and shaping each other, interacting in ways that we are only just beginning to understand. Space itself – that seemingly bare and featureless void – is not as empty as we had thought.

CHAPTER THIRTEEN

RAIN FROM THE UNIVERSE

Thou canst not stir a flower
Without troubling of a star.

    – Francis Thompson[1]

In this day of million-dollar computers and multi-million-dollar accelerators and billion-dollar interplanetary probes, it is useful to remember that some of the most important scientific discoveries have been made using equipment such as a kitchen pan, a cork, and a glass jar. One of the simplest scientific instruments ever invented was responsible for uncovering a scientific mystery that has not yet been solved.

The little instrument called an electroscope consists of a piece of gold foil hanging from the side of a metal support rod in a protected space such as a tightly stoppered glass flask. The metal support, which extends up through the cork, can be given an electrostatic charge by touching it with a piece of rubber that has been rubbed against wool or a glass rod that has been rubbed with silk. The charge distributes itself throughout the metal rod and the gold foil, which is light and flexible enough to move easily. Because the foil has a charge of the same sign as the rod, it is repelled by it and stands out away from the rod at an angle which depends upon the amount of charge involved. Under normal circumstances air does not conduct electricity, so there is no place for the charge to go; theoretically, the little piece of foil should continue to stand out stiffly at this angle forever. But when an electroscope charged in this way sits idle for a while, the gold leaf begins to droop and slowly returns to its original collapsed position. The electroscope has lost its charge. Where has it gone?

Physicists in the nineteenth century who observed this phenomenon had a plausible explanation – the cork was not a good enough insulator. They designed electroscopes that had more perfect insulation, but still the charge leaked away. Although this fact was puzzling, no one really bothered about it; it did not seem very significant at the time.

Then at the end of the nineteenth century Wilhelm Roentgen discovered X rays and he demonstrated that when these powerful rays shone on an electroscope it lost its charge very quickly. A year later Antoine Henri Becquerel found that radioactive elements like radium, when held close to the electroscope, also caused a rapid discharge. X radiation and gamma rays produced by radioactive substances penetrated easily through the glass flask and ionized the air molecules inside the 'protected' space. Ions conduct electricity; they carried the charge away from the foil.

Perhaps, then, the very slow but inevitable discharge of electroscopes was caused by a small amount of X radiation and radioactivity present in the atmosphere. X rays, however, did not appear to be a normal component of our environment; X-ray films remained unclouded after months of storage. Radioactivity was a possible cause. It was known that the earth contained deposits of uranium and thorium, which decayed into other radioactive elements such as the gas *radon*, which emanates continuously from the earth and constitutes a small but ever present fraction of the earth's atmosphere.

This explanation, which seemed to be quite acceptable at the time, was shown to be inadequate just a few years later. In 1900 two scientists in Germany (Johann Elster and Hans Geitel) and C. T. R. Wilson in England conducted a series of experiments which demonstrated that when an electroscope was shielded by walls of lead thick enough to prevent the transmission of X rays and gamma rays emitted by radon, the rate of discharge did slow down but no amount of shielding stopped it entirely. The mysterious radiation that discharged electroscopes possessed extraordinary powers of penetration.

In spite of these facts most scientists continued to believe that the radiations came entirely from the earth; there did not seem to be any other logical explanation. To settle the point some enterprising physicists carried electroscopes on ocean voyages, but they were surprised to discover that the electroscopes discharged just as rapidly when several miles of water separated them from the earth's surface. In 1910 T. Wulf, a German physicist, took his electroscope to the top of the Eiffel Tower (then the highest building in the world) and observed that there was little or no decrease in the rate of discharge. Later the same year a Swiss scientist, A. Göckel, carried the principle of Wulf's experiment a step further. He put even more distance between the earth's surface and his electroscope by taking it aboard a balloon flight. At an altitude of 13,000 feet he found no difference in the rate at which the little gold leaf collapsed.

All of these experiments indicated that there was something wrong with the hypothesis that the earth was the only source of discharging radiation. Scientists, however, are always reluctant to give up a theory unless there is a better one to take its place. Even when a new theory does appear it is not always given a cordial reception.

In August 1912 an Austrian physicist, V. F. Hess, repeated Göckel's experiment and got more interesting results. He ascended to 16,000 feet in a balloon carrying three electroscopes, two of which were enclosed in airtight outer containers to maintain a constant internal air pressure. Hess was accompanied by two assistants, who helped record altitude and temperature as well as the rate of discharge of the electroscopes. Up to approximately 2,000 feet the rate of discharge slowed down, indicating that the earth was the source of at least some of the penetrating radiation. But then above 2,000 feet the rate of discharge began to increase again and this rate accelerated with altitude until at 16,000 feet the electroscopes were losing their charge four times faster than they had at sea level. This was indeed a strange discovery. It seemed to Hess as though the rising balloon must be travelling *towards* a source of radiation. He reached the daring conclusion that very penetrating rays were coming from outer space. This revolutionary proposal met with a solid wall of criticism from other physicists.

In the years between 1913 and 1919 a German scientist, W. Kolhörster, made a number of balloon flights which confirmed the measurements made by Hess up to 16,000 feet. Above that height the effect was even more dramatic. At 29,000 feet the discharge occurred twelve times more rapidly than at ground level. [2]

The American physicist Robert A. Millikan was among the many scientists who found it difficult to accept the strange idea that these very penetrating radiations are bombarding the earth's atmosphere from outer space. In 1922 he and a colleague, I. S. Bowen, performed a set of ingenious experiments. Pairs of sounding balloons carried electroscopes fitted with a device which automatically recorded on photographic film the gradual discharge of the electroscopes as well as temperature and pressure data. The same equipment was lowered into high-altitude mountain lakes and bodies of water near sea level.

The results of these experiments left no doubt in Millikan's mind. The rate of discharge increased regularly with increasing altitude and decreased with depth under water. These facts could only be reasonably interpreted on the basis of Hess's theory. Robert Millikan was a very prominent physicist and his support gained acceptance for Hess's work. Millikan named the mysterious radiation *cosmic rays* – 'cosmic' because they came from some unidentified place in the cosmos. The word 'ray' proved to be a misnomer – as we shall see.[3]

During the next few decades the study of these strange visitors from outer space became one of the most active fields of scientific research. The humble little electroscope went on more and more exotic journeys. Measurements were made in gold mines far below the earth's crust and at the bottom of the sea. They were made in balloons and rockets flying high into the stratosphere. The astonishing truth was gradually revealed. Even when shielded by a thousand feet of solid rock and miles of water a small component of ionization could still be observed. This fact demonstrated that these cosmic rays had almost unbelievable characteristics.

The penetrating power of radiation depends upon its energy. The more energetic the radiation, the more easily it can pass unaffected through a layer of matter. But even though X rays and gamma rays are extremely energetic radiations, they do not pass through a thousand feet of rock.

The measurements made at elevations above most of the atmosphere showed that the intensity of the cosmic radiation increased rapidly with altitude and changed in character. The number of incident rays diminished but the energy of the remaining ones increased enormously. It soon became apparent that cosmic rays – at least the 'primary ones' that strike the upper layer of the earth's atmosphere – are not a form of radiation, after all, but fragments of matter – ions rocketing in from outer space with velocities exceeding anything previously known.

These primary rays appeared to be somewhat deflected by the earth's magnetic field. Like the solar wind, their incidence was higher at the polar regions than at the equator. However, beyond the magnetosphere the distribution of the primary rays proved to be almost completely independent of direction. This fact was one of the major snags in trying to identify the source of the mysterious rays. If they came from the sun, for instance, why was there no directional factor such as we observe in the solar wind? Cosmic rays occurred with as much strength and frequency at night as they did in the day, in cloudy weather as well as sunny. There was no significant variation from season to season. Finally, it was observed that during total eclipses of the sun cosmic radiation was not diminished. From these facts most physicists concluded that cosmic particles must come from beyond the solar system.[4]

In fact, the sun appears to act like an efficient housewife sweeping this cosmic debris from its own hearth. After the discovery of the solar wind and the sunspot cycle, it was shown that increases in solar activity result in a *decrease* in cosmic ray intensities. This effect has been observed immediately following a solar flare and also in a regular cyclic pattern following the eleven-year sunspot period. The best explanation offered for this phenomenon is that the pressure of the solar wind deflects the cosmic rays, sweeping them away from the central regions of the solar system. The deflection is greatest at times of high solar activity but it is an important factor at all times. The cosmic ray energy density in the vicinity of the earth may be only about one third as great as it is in the space between the stars.

The simple electroscope was eventually replaced by other instruments that measure cosmic rays more accurately. The Geiger-Müller counter is one of the most widely used tools. This detector responds whenever an ionizing ray above a certain energy level strikes one of the electrodes, permitting an electric pulse to flow and triggering a recording device. With a Geiger counter attached to an audio circuit it is possible to listen to the 'rain' of cosmic particles: pong . . . . . . pong . . . . pong . . . pong . . . pong . . . . pong. The sound continues at its uneven but steady pace. It is an eerie sensation to listen to this persistent patter of rain pelting in upon the earth from all corners of the universe.

The Geiger counter records only the most energetic of the cosmic rays; other meters, such as scintillation counters, have been designed to measure the 'softer' rays. Each impact produces a flash on the screen, and these softer rays occur so frequently that special electronic devices are needed to count them. Something like a thousand cosmic rays strike the human body every minute of the day.[5] However, the rays that strike us here on earth are on the average hundreds of times less energetic than the primary rays which arrive from outer space at the top of our world. Like the roof of a good house, the various layers of the earth's atmosphere act as an almost solid barrier, preventing the primary rays from reaching ground level. As each of these concentrated pellets of power strikes the molecules of the air, its energy is dissipated and scattered, creating a shower of 'secondary rays'. The energy is now divided up among many atomic fragments; as each of these descends into denser air it breaks up again into more showers, becoming a cascade of particles falling towards the earth. A primary ray may generate a hundred secondary rays, which consist of very small particles of matter and gamma radiation. Many of the secondary rays are *mu mesons*, a type of subatomic particle which is about one ninth the size of a hydrogen nucleus and has the strange characteristic of being unable to react strongly with the nuclei of atoms. This unsociable personality as well as its small size makes the mu meson able to penetrate a deep layer of matter without being absorbed or reflected by it (just as a thin, silent guest might be able to drift through a crowd of people and out the front door without anyone noticing that he had been at the party). Some very fast mesons are able to bore their way through miles of water and bury themselves deep in the mud at the bottom of the sea.

The primary ray particles are single nuclei of atoms stripped of their electrons. Hydrogen nuclei (protons) make up the largest portion – about 90 per cent. Helium nuclei comprise 9 per cent of the total, and the remaining 1 per cent is divided among the nuclei of all the other elements which are normal constituents of matter. In fact, the relative proportion of atoms present in primary rays tells us the proportions in which these elements are present in the universe as a whole. Cosmic rays are a sample, delivered to us here on earth, of the raw material of the universe – material that spins out spiral galaxies like Andromeda and the Milky Way, material that builds red giants and white dwarfs, pulsars and black holes, material for billions of planets and satellites like our moon, material even for mother-of-pearl clouds and salt spray and the winds that riffle restlessly across the face of the sea.[6]

Many wonderful facts have been learned about cosmic rays, but the most tantalizing questions remain unanswered. Where do they come from and what is the source of the enormous energy they possess? Some of the cosmic particles which bombard our upper atmosphere are travelling with almost the speed of light. They are a hundred million times more powerful than the most energetic atomic particle ever produced in the largest accelerators built by man. No one has been able to postulate a completely satisfactory explanation for such incredible concentrations of energy in space, although a wide range of imaginative possibilities has been proposed.

Particles of ionized gases could be accelerated by the complex electromagnetic fields which exist in our galaxy. Although most ions would be slowed down by these fields as often as they were accelerated, a lucky few (like winners of sweepstakes) would happen to strike the fields just right and would be speeded up to colossal velocities. The less lucky slower majority would join the galactic dust cloud.[7]

Other likely sources for the cosmic rays are the stellar explosions called supernovae that occur in our galaxy about once every thirty years. The violence of these cataclysmic events projects very energetic particles and radiation into interstellar space. A third possibility is that the core of our galaxy may be generating cosmic rays.[8]

Physicist John A. Simpson and his colleagues at the Enrico Fermi Institute for Nuclear Studies at the University of Chicago have found evidence that the cosmic rays in the interstellar space of the galaxy have lifetimes on the order of ten million years. But studies of lunar samples returned from the Apollo missions show that the flux of cosmic rays has probably remained essentially constant for at least two billion years. Therefore some mechanism for continuously producing these enormously energetic particles must exist in the great spiral arms that make up the Milky Way.[9]

Once formed, the magnetic field of the galaxy could trap all but the most energetic speeding particles in a giant halo around the Milky Way just as the solar wind is trapped in the Van Allen Belts around the earth. Some of them might leak out of the magnetic trap in the same way that charged particles escape from the radiation belts into the earth's ionosphere. But firm proof that any one of these mechanisms completely accounts for cosmic rays is still lacking.

If our galaxy generates cosmic rays, what about the other galaxies? There is very good evidence that all galaxies contain a flux of cosmic rays. Some appear to have more than our galaxy does. Perhaps cosmic rays are also generated in intergalactic space. Perhaps they pervade the whole universe.

While astrophysicists stretch their imaginations to answer these questions, biologists are concerning themselves with problems closer to home. How do cosmic rays affect life on earth? Do they constitute an important hazard for astronauts and future space travellers?

It is known that a small fraction of cosmic ray energy arrives at the earth's surface in a form concentrated enough to do biological damage, disrupting cells and causing alterations in the structure of genes, the hereditary units that transmit characteristics from generation to generation. A change in the gene molecule produces a mutation, which is almost always deleterious. Small disabilities vary all the way from allergies to albinism. More important mutations cause gross abnormalities such as mental retardation and mongoloidism. Two out of every hundred babies born today have a serious genetic defect. On a worldwide basis this mutation rate results in 1,600,000 defective births every year. Mutation in body cells is also known to be one of the causes of cancer. Some of these human tragedies can certainly be attributed to cosmic radiation, but no one is sure just how many. It is estimated that cosmic radiation accounts for about one fourth of the background radiation to which we are all exposed at sea level. The other three fourths is due to radioactive elements originating in the earth. However, mutations are also caused by chemicals in the food we eat and the air we breathe.[10]

One of the most interesting discoveries about cosmic rays in recent years involves their production of a radioactive element in the atmosphere, carbon 14. The radiocarbon atoms enter into chemical combination; they are taken up along with normal carbon atoms by all plants and animals, and used to build organic molecules. Radioactive elements decay gradually over a period of time. After 5,000 years slightly less than half of the carbon 14 originally present is transmuted into other elements. Therefore, assuming that the concentration of carbon 14 in the atmosphere remained essentially constant, the age of any carbon compounds derived from atmospheric carbon can be determined by measuring the radioactivity left in the sample. This dating method or 'clock' has been widely used.

But now it has been discovered that the concentration of carbon 14 varies in a way which is directly related to the flux of cosmic rays striking the top of the earth's atmosphere. During periods of high solar activity there is a smaller component of carbon 14 in the atmosphere because the more intense solar wind sweeps a larger proportion of the cosmic rays away from the solar system. When the solar wind is weak more cosmic rays get through, strike the molecules of the air, and create carbon 14. Cyclic variations corresponding with the eleven-year sunspot cycle have been measured in tree-ring studies, and unusually large amounts of radiocarbon have been found to coincide with long periods of reduced solar activity, such as the years between 1645 and 1715. The concentration of radiocarbon in the air during these periods appears to have been as much as 25 per cent greater than at times of high solar activity.[11]

These discoveries have necessitated a correction in the radiocarbon dating scale and they raise significant questions about the effect of higher cosmic ray flux on living organisms. Carbon 14, for example, is known to cause genetic mutations, a greater incidence of defective offspring and neonatal death. While exposure to cosmic radiation on earth varies only slightly because of the protecting layer of the earth's atmosphere, in space the exposure is many times greater than it is on the earth's surface. Biological specimens taken along on space missions have shown increased mutation rates, but more definite long-term experiments are needed before it will be possible to estimate the biological hazard to astronauts from cosmic rays in space.

During the first three weeks of the Skylab 4 mission, Edward Gibson and William Pogue reported observing occasional bursts of intense light-flash activity which lasted for about 5 to 10 minutes. These occurred during sleep periods when the men were lying awake in darkened compartments. It is believed that the astronauts were actually 'seeing' cosmic rays, that the flashes were caused by particles passing through the eye after it had become adapted to the dark. Regular planned observation periods were then set up using light-tight goggles. The men reported each sighting, which usually took the form of a short streak or 'tadpole' with a spotlike head at the leading end. Occasionally a round or cloud-shaped flash was seen. During these observation periods (which lasted for 70 and 55 minutes) scientific instruments aboard Skylab reported the incidence of cosmic rays. When the two sets of data were compared there was a very strong correlation, confirming the theory that the visual flashes were caused by cosmic rays.[12] These were probably secondary rays, for the metal shell of the satellite provides at least a partial shield against the primary particles. As the primary rays strike the satellite they give birth to a cascade of small particles of matter and pulses of very penetrating radiation which pass through the visors into the eyes of the astronauts. The speeding nuclei of the primary rays themselves are so packed with power that the fastest of them could cause a flash in a small electric light.[13]

During space walks and the exploration of the moon when astronauts left the protection of their spaceship, the primary rays fell directly on their clothing. The helmets worn by the men who walked on the moon were found to be scarred by microscopic tracks caused by the impact of primary cosmic ray particles.[14] However, because of the short length of time involved, the amount of cosmic radiation to which these men were exposed is believed to be small. Continuing studies will throw more light on the question of the radiation hazard associated with longer space missions.

The bombardment of speeding cosmic particles on the helmets and the sheath of a satellite makes no noise, but astronauts could hear the rain of another type of cosmic matter on their spacecraft. Tiny meteoroids about half the size of a grain of sand make a little ping 'like birdshot pellets' when they strike the shell of the satellite.[15] In order to study this phenomenon scientists attached samples of material to long-orbiting spacecraft. When these were collected later they were found to be pockmarked with small dents made by the micro-meteoroids. The scientists estimated that on the average one of these tiny grains of matter strikes a surface one yard square every hundred seconds in space. The chances that a meteor as large as a garden pea would hit – and perhaps pierce the skin of a satellite – during a month are about one in a million.

Meteors are pieces of matter from interplanetary space or dust from comets' tails. Formed from the original solar dust cloud at the same time the planets were born, countless numbers of these miniature satellites have rotated like the planets in orbits around the sun for billions of years. Colliding occasionally with their neighbours, bombarded by cosmic rays and solar wind, they have been broken up, scarred, and eroded in their long journeys through space. A few are attracted to and join larger particles. Beyond the orbit of Mars lies a band of asteroids – sizeable chunks of rock weighing anywhere from a pound to many tons. Some meteoroids may be chips from these asteroids and some may be dust from interstellar space gathered up by comets that swing in elongated elliptical orbits far beyond the solar system.

The presence of this cloud of matter can be observed on earth in the phenomenon known as *zodiacal light*, a pale ghostly beam of illumination that can be seen in the sky on moonless nights over the ocean or in the country far away from distracting city lights. In the middle latitudes the most favourable times to observe zodiacal light are January, February, and March in the western sky after sunset. About two hours after the sun has gone down a very faint wedge-shaped cone of light rises in the sky; it is tipped obliquely towards the south-west in the Northern Hemisphere. As the night progresses the cone grows larger and becomes more upright – an enormous pyramid of very diffuse light, approximately as bright as the Milky Way. Near midnight the western zodiacal light begins to fade and later the same effect appears in the east, reaching its maximum brightness several hours before sunrise. The path of these pyramids of light lies along the zodiac, the apparent course of the sun across the constellations.

At first the source of zodiacal light was thought to be sunlight reflected on a relatively small cloud of dust particles immediately surrounding the sun. Like motes of dust on earth, these become visible when they are illuminated by a beam of light after the disk of the sun itself is hidden by the earth.[16]

When the information collected from the early scientific satellites was analysed, it became apparent that this dust cloud extended at least as far as the earth's orbit itself. Then in 1972 and 1973 the Pioneer missions to Jupiter and the Helios solar probes carried light meters that recorded zodiacal light much farther into space. Such measurements confirmed the fact that the cloud of dust reaches out beyond the orbit of Jupiter. Now as Pioneer 11 races at 26,000 miles an hour towards a rendezvous with Saturn in 1979, messages come trickling back to us across the expanding track. 'That veil of matter is growing thinner but it is still here far beyond Jupiter, 450 million miles from the sun!'[17]

As the earth swings in its orbit through interplanetary space it intercepts vast numbers of oddly assorted pieces of matter. Accelerated by the earth's gravitational field, they are travelling very rapidly when they strike the atmosphere and collide with air molecules. The force of these collisions converts the solid pieces into a cascade of charged particles that become incandescent. Like miniature comets, these shooting stars have tails that remain luminous long after the meteoroid has disintegrated.

Gordon Cooper described a meteor shower seen on his first orbital flight in 1963: '. . . the most fascinating things,' he said, 'were meteoroids striking the earth's atmosphere. I could just sit there and look down on them coming in from all directions – big ones and little ones, in short streaks and long streaks. Once I saw a big one that came in to the side of the earth and seemed to be heading straight up toward me. Then it appeared to burn out.'[18]

Every day more than a million billion meteoroids impact on the roof of our atmosphere. Usually they burn out long before they reach the surface, leaving nothing but ashes and dust to float slowly earthwards. Scientists say that several tons of meteor dust enter the earth's atmosphere every day, enough to double the weight of the planet by the year 3,000 million million A.D. (This is a long time even on a geological scale; the earth would be almost a million times older than it is today.)

Some meteorologists believe that the presence of this dust in our atmosphere affects the earth's rainfall. E. G. Bowen, an Australian scientist, is one of the principal proponents of this view. He bases his theory on records of rainfall at seven different control points throughout the world. These showed that persistent rains occurred usually about thirty days following the passage of the earth through a meteoric shower. The delayed reaction is due, Bowen says, to the time required for meteor dust to travel from fifty- or sixty-mile elevations to the troposphere.[20]

Occasionally a meteor large enough to survive the trip through the atmosphere falls to earth. As it approaches the planet it is moving with a velocity of 8 to 45 miles per second and when it hits the ground it may dig a deep crater and scatter meteoritic fragments over the surrounding terrain.

One of the most spectacular recent falls occurred in Norton County, Kansas, in 1948. It was a sunny February day with a blue sky overhead. At about five o'clock in the afternoon in the small town of Jennings, eleven-year-old Creta Carter was taking the family laundry down from a clothesline in her back yard. Suddenly a brilliant ball of fire blossomed out in the clear sky, flashing directly across her field of view. Several detonations of sound followed rapidly one upon another like a cannonade, and the fireball turned into a red streak followed by an angry boiling cloud. The air was filled with hissing sounds. Undismayed by this frightening apparition, Creta calmly watched the smoking mass fall and marked the place where it disappeared behind the town's largest building.

Although the ball of fire was bright enough to be seen for hundreds of miles, Creta was one of the few people to have her face tipped up to the sky at exactly the instant when the object burst into flame. Moments later thousands of people rushed out of their homes, alerted by the loud noise, and looked up to see the mushrooming cloud. Most of them concluded that an atomic bomb or a rocket had been detonated over their quiet rural community. But experts who were called in to solve the mystery soon discovered that an unusually large stony meteor had fallen on Norton County. It had caught fire at about thirty-five miles above the earth. The intense heat had caused the meteor to explode, pouring out trails of smoke clouds and many fragments which whizzed through the air, making a bedlam of strange sounds. The pieces were scattered over many miles; the largest one weighed over 2,000 pounds and penetrated eleven feet into the ground.[21]

Larger meteors than this have impacted the earth's surface at earlier times in the planet's history. Craters up to a mile in diameter have been

28. Large meteor crater three miles in circumference near Winslow, Arizona.

found. One very spectacular event occurred in Siberia in 1908, causing forest fires and pressure-wave damage over many square miles. Mysteriously, no crater was produced and no fragments have been found. Since the fall occurred in late June and the soil in that region becomes a swamp during the summer months, it is possible that the fragments penetrated so deep that they have not yet been discovered.[22]

Comparatively speaking, however, large meteorite falls are rare on earth. The moon, Mercury, and Mars, with little or no atmosphere to protect them, are deeply pitted with many meteorite craters, but on earth only one to ten meteorites reach the surface every day. Most of these are small, weighing just a pound or two. The vast majority are ignited and scattered by our atmosphere. They are turned into a soft shower of dust, a bright display of shooting stars in our night sky.

Thus earth's aura transforms the matter and energy that rains on us day after day from the whole universe – matter that has been blown by the solar wind off a comet's tail, or was spewed into space by an exploding star across the Milky Way, or bred in the farthest reaches of the universe where objects more luminous than a hundred galaxies are pulsing with an unknown source of power. The destructive force of this matter that arrives at the outer boundaries of our world is rendered almost entirely harmless by the atmosphere, shaken up and divided into benign little packages that can be absorbed and utilized by the fragile assemblies of living matter that inhabit the earth.

Endlessly, day after day, this gentle rain of cosmic matter drifts down through the many layers of air. When it reaches the troposphere terrestrial rain speeds its journey, washing it down into the Coral Sea, onto the wheat fields of Montana and the tea terraces of Luzon. A ladyslipper blooming on the forest floor may cup a drop of this cosmic solution for an hour before the vagrant breezes shake it out, scattering it onto the mat of last year's leaves, where it trickles slowly down and touches at last the body of the earth itself.

CHAPTER FOURTEEN

ATMOSPHERES OF THE SOLAR SYSTEM

We are the music-makers,
We are the dreamers of dreams . . .
Yet we are the movers and shakers
Of the world forever, it seems.

– Arthur O'Shaughnessy[1]

It was four o'clock on an early summer afternoon. The dim sun had swung low towards the western horizon, casting long jagged shadows of rocks and sharp escarpments across the rust-red landscape. Suddenly a tiny cloud blossomed out in the pale pink sky. It grew rapidly larger and a few seconds later a spiderlike metallic object unfolded itself beneath the cloud and extended three round feet on spindly legs. The cloud crumpled and disappeared; the spiny object exhaled sharply, sending billows of red dust up from the rocky surface. Then it settled down with a slight jolt in the wide basin of Chryse Planitia. Viking 1 had landed safely; it was 20 July 1976, earth-time.[2] And the dust had not yet settled when the robot started its inspection of the Martian surface, sending up radio signals that would travel 212 million miles back to earth.

Within a few days detailed colour photographs had been relayed, combined, enhanced with the aid of computers, and published around the world. These pictures contrasted sharply with photographs of the moon, which had shown a stark grey and white landscape against a black sky. The fact that Mars has a daytime sky tells quite a bit about the planet. It has an atmosphere which scatters and diffuses the sunlight that falls upon it. Clouds are sometimes seen in it and morning haze. The faint pinkish colour is believed to be caused by dust in the air.

Eight days after its perfectly executed landing, Viking 1 extended a long mechanical arm, scooped up a sample of the dusty Martian soil, and began a series of tests for the presence of life. These tests, performed under the most alien conditions, were remarkably similar to the experiments that Joseph Priestley had performed 200 years earlier in a castle in England.

Priestley, you remember, put a mouse and a sprig of mint inside enclosed volumes of air and found that the characteristics of the air were altered by the presence of the animal or the plant. Now, if his experiments had been conducted blind – if Priestley had not been able to see the mouse or the mint – their presence could have been inferred from the change in the chemistry of the gases inside the closed containers. The mint plant would have produced oxygen as a waste product of photosynthesis and gradually used up the carbon dioxide in the vessel. The mouse, on the other hand, would use up oxygen and cause an increase in the amount of carbon dioxide in the sample. In either case something would be happening to the chemistry of the gases, a continuous change caused by the vital activity of the animal or the plant. If the experiment were passed through an oven and sterilized at temperatures high enough to kill the mouse and the mint, then the change in the chemistry of the sample would cease.

The Viking experiments – although more sophisticated in detail – were based on these same principles. But confusing results were obtained from the tests. Some of them seemed to indicate the presence of living things, but others were most readily explained as the result of inorganic reactions when the Martian soil was combined with the nutrients and gases used in the experiments. An increase in oxygen did occur in the experiment designed to test that reaction – a much larger increase than typical micro-organisms on earth would have produced. Carbon dioxide (tagged with radiocarbon) also increased dramatically in one of the experiments. Both of these activities, however, levelled off and finally dwindled to a stand-still within a few days. If these changes were caused by living organisms, the rates of their vital activities must be quite different from the rates characteristic of organisms on earth. Perhaps the processes took place much faster and used up the ingredients introduced into the experiments. Or perhaps the chemical reactions were inorganic and, therefore, could not proceed continuously.

Another interesting fact was discovered when the sample of Martian soil was subjected to high temperatures. The activity sharply decreased, as would be expected if living organisms had been killed by the heat. On the other hand, an instrument for measuring the concentration of the various types of chemicals reported that a lot of water was present in the soil but there was no sign of organic molecules. Any micro-organisms that

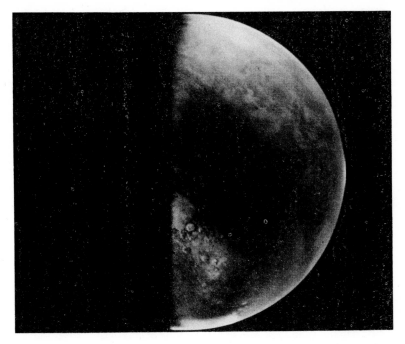

29. *A composite photograph of Mars taken by Viking 1 Orbiter on 18 June 1976.*

may be present do not appear to leave measurable amounts of organic remains in the soil when they die. The complete set of data from these experiments could not be simply explained in terms of either chemical or biological processes as they are known on earth.[3]

A third possibility – and a much more exciting one – also exists. If life is found on Mars it will be the first example of a vital activity that evolved in complete isolation from life on earth. There is no reason to believe that Martian organisms would photosynthesize or metabolize in just the same way or at the same rates as terrestrial micro-organisms. Even the building blocks of living matter might be somewhat different. In fact, the chances of a perfect match are considered to be very small indeed. So the ambiguities in the experiments may foretell a more important finding than if they had exactly followed the expected pattern. This discovery could open up a whole new perspective on the nature of life.

If even the most primitive forms of living things are found on the rocky red planet, this fact would demonstrate that life is not just a chance aberration which occurred once in the history of the universe but that it is inherent in the cosmic process. It would also demonstrate an extraordinary degree of toughness and tenacity in life processes. Mars, as surveyed not only by the Viking satellites but also by earlier missions, is certainly not an ideal cradle for life as we know it. It is surrounded by an atmosphere so thin that the pressure at the planet's surface is just $^1/_{144}$ of the pressure on earth. The Martian air is approximately 95 per cent carbon dioxide, 3 per cent nitrogen, 1 to 2 per cent rare gases, and only 0.3 per cent oxygen. It contains very little water vapour and most of the ultra-violet component of sunlight penetrates to the surface. Temperatures range between $-32°F$ ($-36°C$) on the warmest summer days and $-127°F$ ($-88°C$) in the polar night. All in all, the planet seems to possess an inhospitable environment made even more forbidding by the presence of high winds.

The regular occurrence of dust storms on Mars has been inferred from telescopic observations on earth. Small local storms seem to appear frequently, and once every Martian year an enormous storm is observed involving at least half and sometimes the whole planet. (Mariner 9 arrived in the vicinity of Mars in the middle of such a storm.) At the end of the Martian spring, when the planet is at its closest approach to the sun, a shining white trail appears in the southern hemisphere. It extends for thousands of miles beginning about 30° south latitude. After a few days the trail becomes yellow or reddish in colour. It widens and spreads rapidly, moving in a westerly direction. Soon it circles the globe and, extending southwards, it envelops the entire southern hemisphere. Occasionally – as in the storm witnessed by Mariner 9 – it spills over into the northern hemisphere and within a few weeks it has encompassed the whole planet.

It is theorized that these storms begin and grow like hurricanes on earth. When ground temperatures reach a maximum, surface winds rise strongly enough to carry dust high into the air. The planet's rotation imparts a circular component to these winds. (Mars turns on its axis once in 24.5 hours, its day being approximately the same length as ours.) The dust in the air absorbs energy from sunlight and heats the air further; thus a self-augmenting process develops in a way which is analogous to terrestrial hurricanes, in which columns of rising hot air are further heated by the condensation of water vapour. On the edge of the Martian storm the strong temperature contrasts accelerate the winds and the storm grows until a whole hemisphere or the whole globe is covered. Then the temperature differences decline; the winds gradually subside; and the dust settles back to the crust of the planet. Wind velocities in these storms must exceed a hundred miles an hour to carry sand to these altitudes and, as Mariner 9 reported, visibility is totally obscured for weeks. Such a storm would make the haboobs of the Sudan seem like a clean gentle breeze on a summer's day.[4]

Mariner 9, however, stayed in orbit long enough to outlast the storm and obtain spectacular pictures of a strangely variegated landscape with enormous volcanic craters and deep canyons which were probably caused by movements of the Martian surface. In addition the photographs revealed a curious network of large channels that looked exactly as though they had been formed by running water. The more detailed photographs taken on the Viking missions lend further support to this hypothesis.

The evidence now seems strong that water in the liquid form existed on Mars at least once and perhaps many times in its history. But where is the water today? It is possible that large quantities may be bound up in the form of subsurface ice layers like the permafrost in the polar regions on earth. A vast reservoir of water is known to be stored in the glistening white polar caps. These glaciers are augmented during the winter season in each hemisphere by large deposits of frozen carbon dioxide which evaporate again every summer. Perhaps at some time, the scientists reason, the whole planet was warmer. Then the permafrost and polar caps would have melted or evaporated, releasing water and carbon dioxide. As these entered the atmosphere they would have helped to hold the heat of the planet and moderate the temperature differences. A more stable climate would have developed, with rains and water from the melting glaciers pouring down onto the soft, dusty land. Unprotected by vegetation, this loose soil would have been cut into deep canyons and arroyas resembling those cut by flash floods in terrestrial deserts. In this milder environment life could have evolved and, as the climate deteriorated, some organisms might have adapted to the more rigorous conditions.

The most obvious reason for such a dramatic climatic change would be a significant alteration in the amount of energy received from the sun.

Astronomers calculate that a fluctuation of about 10 to 15 per cent in solar radiation would be sufficient to account for such a change in the environment of Mars, and they speculate that the latest change may have occurred millions, perhaps even billions of years ago.[5]

If the sun's energy has been variable during the lifetime of Mars, the climate of the other planets must also have been affected. On earth there have been major shifts in average world temperature. About 250 million years ago glacial ice sheets covered regions that are now just 10 or 20 degrees from the equator. Following this glacial epoch a long warming trend peaked about 135 million years ago. Trees and ferns typical of a mild climate flourished around the world. Dinosaurs inhabited every continent from northern Asia to the southern tip of Africa. A few million years ago another ice epoch descended on the earth. These facts are well known, but causes other than changes in solar radiation have been advanced to explain them. Continental drift has brought about a rearrangement of the earth's land masses; for example, regions now at the equator may earlier have been much nearer the poles. This hypothesis, however, does not seem adequate to explain the periods of geological history when the average temperatures around the whole planet were decidedly warmer or colder than they are today.

Another favoured explanation is a change in the tilt of the earth's axis, producing cyclic variations in the average solar energy received on various portions of the earth's surface. These cycles, however, repeat in thousands instead of millions of years. The interval of 250 million years between major glacial epochs on earth raises the interesting possibility that the cycle is related to the time it takes for the whole solar system to swing in its enormous orbit around the Galaxy. W. H. McCrea, an English astronomer, has recently revived a theory originally proposed by Fred Hoyle and R. A. Lyttleton. They suggested that in its trip around the Milky Way the sun and its little flock of planets may encounter varying amounts of interstellar dust that could cause changes in the total amount of solar energy reaching the earth. Seen in the light of this theory, the passage from the ice age of the early Permian period to the dense jungles of the late Jurassic period would be simply seasonal changes in the galactic year.[6]

Although astronomers debate the validity of this theory, its credibility has been reinforced by the discovery of the mysterious dry riverbeds on

Mars. A deeper understanding of our own planet's history is emerging from space exploration. Every day more evidence of the development of the solar system and earth's place in that drama continues to pour in from deep space probes and missions to other planets.

Venus was the first planet visited by a satellite from earth. On 12 March 1961, the Russians launched a probe which passed within 60,000 miles of earth's closest neighbour in space, but unfortunately all radio contact with the satellite was lost before it reached the vicinity of Venus. The next year a U.S. probe, Mariner 2, passed within 35,000 miles of Venus. Scanning the disk of the planet, it measured temperatures as high as 800°F (about 430°C). This information confirmed earlier measurements made by radio telescopes on earth, but since neither of these experiments had penetrated to the surface of Venus, some scientists still clung to the hope that temperatures on the planet itself might be more moderate. Although Venus intercepts twice as much sunlight as the earth does, these scientists surmised that the pale yellow clouds shrouding the planet might reflect a large part of this energy, and the highest temperatures might occur within the clouds themselves rather than on the planet's crust.

Later exploration, however, destroyed any hope of finding life on Venus. Between 1975 and 1979 several Russian and American missions sent back an astonishing story of conditions on the planet that had once appeared to be the nearest twin to the earth in the solar system.

Photographs of the Venus landscape show old mountain formations, rocks, and shallow craters. Surface temperatures ranging up to 720°F (400°C) have been recorded, hot enough to melt lead. Venus's thick atmosphere, which is 90 times as heavy as ours, is composed almost entirely of carbon dioxide with minor amounts of other gases such as nitrogen (3 per cent), sulphur dioxide and water vapour (0.1 per cent). The clouds consist principally of concentrated sulphuric acid, a very toxic and reactive chemical. The hot rain which falls from the heavy yellow clouds of Venus may be the most corrosive fluid in the solar system.

The intense heat on Venus can be explained now that we understand the composition of the atmosphere. Carbon dioxide has the property of transmitting radiation in the visible portion of the spectrum, so most of the sunlight passes through it unhindered. But when this radiation is

30. Venus as photographed by Mariner 10.

intercepted by solid materials on the planet's surface, it is changed to long-wave infra-red radiation. Carbon dioxide does not transmit most of these longer waves; it absorbs and re-radiates them. Thus it acts as a filter, allowing the sun's energy to come in and only a small proportion of it to pass back out again into space. As temperatures build up, any carbon dioxide present in the rocks and soil is 'sweated out' and joins the gases of the atmosphere, increasing the concentration of carbon dioxide and the effectiveness of the filtering action, which is often referred to as 'the greenhouse effect'. Temperatures rise higher and this positive feedback process continues until all the carbon dioxide has been removed from the crust of the planet.

Very strong winds sweep through the clouds of Venus's atmosphere. Winds with velocities over two hundred miles an hour have been observed, comparable to jet streams on earth but extending over much broader regions on Venus. The presence of these powerful air currents has not been explained. The planet rotates very slowly – sunrise and sunset occur only once in every 117 earth-days.

Oxygen in the gaseous state has not been found on Venus, and water – that essential medium of life as we know it – is almost entirely missing. Did Venus ever possess water in quantities comparable to the amounts on earth or Mars? If so, where has it gone? On a planet where surface temperatures are hot enough to melt lead, water cannot be bound up in polar ice caps or permafrost as it is on Mars. Scientists speculate that Venus may have received a smaller endowment of water in its original make-up, and what little water it did receive is almost entirely bound up in the sulphuric acid of the dense yellow clouds. [7]

The conditions on Venus, which might serve as a model for a modern version of Dante's Inferno, offer an astonishing contrast to the environment on Mercury, the little planet which is Venus's next-door neighbour on the other side, occupying the orbit closest to the sun. Mercury was surveyed by Mariner 10, which passed near it three times in 1974 and 1975. Photographs showed a rocky cratered surface very much like the moon. Mercury has almost no atmosphere. This finding was no surprise; Mercury is so small that its gravitational field is not strong enough to prevent gases from drifting away into space. Because it receives the largest influx of solar energy and has no atmosphere or surface water to temper the climate, Mercury has the most extreme temperature range of any planet in the solar system. Noontime temperatures at the equator soar to 800°F (about 430°C) and the dark side cools down to about − 300°F (about − 180°C). Night and day are very long on Mercury; it makes a complete rotation with respect to the sun only once in every 176 earth-days. The polar regions are always well below the freezing temperature of water, and scientists wonder whether some water may be bound up in the form of ice deposits beneath the surface. So far, however, no evidence of this has been observed. Mercury may be a waterless planet.[8]

The big surprise was that Mercury has a magnetic field. This information was totally unexpected and remains unexplained. If – as presently theorized – the magnetic field of a planet arises from electric currents associated with the fluid motions in the core of the spinning planet, why should Mercury, which rotates so very slowly, have a magnetic field much stronger than that of Mars, which spins almost as fast as the earth? The discovery of these facts has demonstrated the need for a new theory of planetary magnetism.[9]

Far beyond the orbit of Mars, beyond the band where the dark, lifeless asteroids spin their endless web of motion, are the spectacular outer planets. Jupiter, Saturn, Uranus, and Neptune are remarkable for their size and their low density in comparison with the little inner planets (Mercury, Venus, Earth, and Mars), which are composed of more tightly compacted material.

Jupiter is the most conspicuous and by far the largest planet in the solar system. Its mass is almost two and a half times that of all the other planets put together. In many ways Jupiter resembles the sun more closely than it does the earth or Mars. Like the sun, Jupiter is composed largely of hydrogen and helium, and these appear to exist in approximately the same proportion as they do on the sun – ten parts of hydrogen to one part of helium. The giant planet radiates about twice as much heat as it receives, but most astronomers believe that this heat is left over from the original condensation of the planet. Some condensation may still be occurring and producing more heat today. However, the giant planet probably never reached the temperature and pressure conditions necessary for the generation of energy in the same manner as the sun and other stars.[10]

Jupiter can be visualized as an enormous drop of liquid hydrogen with a tiny nucleus of iron and silicates. Spinning very fast in space, it makes a complete rotation in slightly less than ten hours. The liquid surface is surrounded by a thick swirl of gases, which have been identified as hydrogen, helium, ammonia, and methane. Some water vapour is also present.

31. Jupiter's red spot can be seen to the right on this photograph taken by Pioneer 10.

When Pioneer 10 and 11 swept past Jupiter in 1973 and 1974, they radioed back the news that the atmospheric pressure on Jupiter is sixty million times the pressure on earth. Any man-made satellites or instruments lowered into such an atmosphere would be crushed like eggshells in an avalanche long before they reached the liquid 'surface'. But actually, there is no well-defined surface on Jupiter; there is a continuous grada-tion between the gaseous and liquid states. As the spacecrafts passed around the planet they measured temperatures within this deep, diffuse 'surface' of about −238°F (−150°C). There was remarkably little differ-ence between day and night temperatures, implying an efficient circula-tion of the atmosphere. Swift jet winds and trade winds similar to those on earth, but much stronger, seem to be typical of the Jovian atmosphere. They create cloud eddy and wave formations that are visible by telescope from earth. Pale yellow, white, and dark rust-coloured bands are present, persisting for long periods of time.

An enormous glowing red spot has been observed for more than three hundred years since Giovanni Domenico Cassini reported it in 1665. Although the spot has darkened and faded frequently during that time, it has never disappeared. It continues to occupy an area larger than the entire surface of the earth, approximately 8,500 miles in width and 18,000 to 24,000 miles in length, beginning at a point slightly south of Jupiter's equator. Information collected by the Pioneer missions points to the conclusion that this Great Red Spot is a gigantic tropical cyclone brewed in Jupiter's 'hurricane belt' and feeding on energy derived from the condensation of water at altitudes somewhat beneath the top of the spot. This Jovian hurricane is a storm which waxes and wanes but never lets up.

One of the most intriguing riddles about Jupiter's atmosphere is the reason for the yellow and dark red colouration. The gases which have been positively identified there are colourless; so other substances must be present to cause these characteristic markings – perhaps phosphorus, as proposed by John Lewis of Massachusetts Institute of Technology – or perhaps organic molecules, as proposed by Carl Sagan of Cornell University. If they are organic molecules, their presence might be an indication that some living organisms exist in a stratum of the Jovian atmosphere where temperature and pressure conditions are within the range that could support life.[11]

Beyond Jupiter in the far-off space of the outer regions of the solar system spin three more giant planets (Saturn, Uranus, and Neptune) and tiny Pluto, the most distant planet of all. It will take Pioneer 11 five years to travel from Jupiter to its rendezvous with Saturn. Beyond Saturn five more years would be required for a satellite to reach Uranus, four to six more for Neptune or Pluto. These still unexplored planets have been studied with instruments based on earth or in orbiting observatories such as Skylab. Pluto is too small and too far away to yield much information in this way, but astronomers do have a tentative picture of the larger planets.

Seen through a telescope Saturn is a beautiful object. Pale gold in colour and brighter than most stars, it is surrounded by shining white rings. Its deep atmosphere is variegated with clouds which are probably composed of ammonia and methane.[12] The rings were at first believed to be formed of ammonia crystals, but more recent information seems to favour the presence of ice crystals of considerable size. Calculations by James Pollack of NASA's Ames Research Center suggest that they are typically five to ten inches in diameter, a circling drift of giant snowflakes.[13] Saturn, like Jupiter, is thought to be mostly composed of hydrogen and helium. It also radiates roughly twice as much energy as it receives from the sun.

Uranus and Neptune, on the other hand, which are smaller and denser than Saturn, contain a much larger percentage of rock and metal and a smaller envelope of liquid hydrogen. They both have remarkably clear atmospheres where sunlight can penetrate deep before it is reflected, and the most abundant gases appear to be hydrogen and methane. Some helium and ammonia are probably also present. Although the amount of include hydrogen – on a body of this size is a mystery. With such a small gravitational field, most of the gases, and especially hydrogen, should rapidly escape into space. One possible explanation is that gases are constantly being replenished by volcanic action. It has been suggested that the atmosphere of Titan contains nitrogen formed from ammonia by the action of solar radiation. If so, the conditions on Titan might be similar to those on the primitive earth. Surface temperatures on Titan are estimated to be −200°F (−128°C); its crust may be composed of a thick

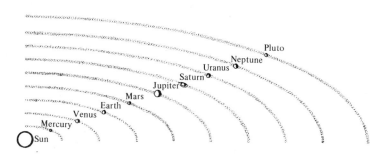

Fig. 10. The solar system: the sizes and distances of the planets are not to scale.

layer of ordinary ice. Everything that we have been able to learn about this satellite of Saturn shows an exceptional body unlike anything else known in the solar system. Astronomers are waiting with high expectations for Pioneer 11 to pass close by Titan in 1979.[15]

The whole field of planetary physics is experiencing a period of explosive growth and intellectual ferment. Unexpected new facts are constantly being received from space.

Carl Sagan expresses the excitement of many of his colleagues who have witnessed this burgeoning of knowledge about the solar system:

> We have entered, almost without noticing it, an age of exploration and discovery unparalleled since the Renaissance, when in just 30 years European man moved across the Western ocean to bring the entire globe within his ken. Our new ocean is beyond that globe: it is the shallow disk of space occupied by the solar system. Our new worlds are the sun, the moon and the planets. In less than 20 years of space exploration we have learned more about those worlds than we have in all the preceding centuries of earthbound observation.[16]

Even a casual reading of these facts reveals an astonishing diversity of atmospheres and environments among our sister planets. From the unprotected surface of Mercury blasted by heat, meteorites, and cosmic rays to the cold poisonous gases that lie with crushing weight on the liquid surface of Saturn as it swims in its aureole of ice – we have surveyed a wide gamut of uninviting possibilities. Only the earth out of all these children of the sun is endowed with the perfect combination of warmth, moisture, and the essential ingredients for life. It is hard to imagine how these planets could all be members of one family. Were they born of the same parents? If so, how did they grow and develop to become so very different?

The answers to these questions still lie in the realm of theory and conjecture; the ideas are changing rapidly as knowledge is collected from space exploration. But scientific thought cannot operate effectively without a theory; there must be a model against which to evaluate new facts and to serve as a guide to suggest new possibilities in the search for knowledge. The most generally accepted explanation of the origin of the solar system is an integral part of the theory of the creation of the universe itself.

It was less than three hundred years ago that Archbishop Ussher dated the creation of the world at 4004 B.C. According to the Bible story, on which this conclusion was based, the earth and the universe were created on the same day. Bishop Ussher gave the day and even the hour: 9 a.m. on Sunday, 23 October. What a satisfying, cosy thought that was! Better than knowing your own birthday down to the hour and minute when you uttered that first cry and were placed in your mother's arms. But this comfortable idea did not long survive scientific scrutiny. By the late nineteenth century Lord Kelvin had pronounced that the world was 22 million years old. In 1932 astronomer Edwin Powell Hubble discovered that the universe seemed to be expanding. Extrapolating backwards from these measurements, he calculated that the expansion must have started 2 billion years ago. This was assumed to be time zero, the beginning of the universe. New measurements made in 1956 caused this estimate to be increased to 5 billion years, and in 1958 it was revised a second time, bringing the age to 10 billion. Like the tantalizing mirage of the oasis in the desert, the birth date of the universe has been receding before us. As we advance in our knowledge, the beginning of the universe retreats ever farther back down the dimly lit corridors of the past. Cosmologists now talk in terms of 18 billion, perhaps even 20 billion years, and they qualify these statements with a measure of humility.[17]

We really don't know exactly *when* the universe was created, but most cosmologists believe that they have a conceptual picture of *how* it was created. They think that it began with a concentration of matter and energy so enormously compacted that all the stuff of the present universe was consolidated into one large lump. At time zero, the lump exploded with a force so great that the pieces are still hurtling away from each other in all directions. The fragments of the original 'egg' were as finely divided as matter can be – not even whole atoms, just pieces of atoms. This intensely hot plasma pushed outwards into space and the random forces of the explosion distributed it into clouds of varying density. Time passed and the plasma cooled; simple atoms like hydrogen were able to form, making up the dust clouds which gradually filled the universal space. In regions of greatest density these clouds began to contract under the force of gravity. They formed whirlpools and eddies. As they continued to contract they rotated faster (just as a figure skater spins faster as she moves her arms in close to her body). From these clouds of collapsing gas a first generation of stars was born, and the gas that was left over from this star formation settled into disks or spheres of rotating particles. As the first stars became more and more compacted by gravity they heated up, and when the internal temperatures reached a critical value, fusion processes began to occur, turning hydrogen into helium with the production of impressive amounts of energy. This process is believed to take place in all stars, including our own sun, and is the source of the enormous amounts of energy they pour forth so prodigally over eons of time.

Fusion of four hydrogen atoms to one helium atom occurs at temperatures of 3,600,000°F (2,000,000°C). As this process continues in very large stars temperatures rise even further and new reactions begin to take place in the core of these stars. At temperatures between 180 and 360 million degrees Fahrenheit (100 and 200 million degrees C) – three helium atoms fuse to make one carbon atom. This also releases large amounts of energy and it is followed at higher temperatures by the fusing of a carbon atom with a helium atom to make oxygen. As the temperatures continue to rise, more reactions occur, creating many of the heavier atoms. Finally, when temperatures of around 5.5 billion degrees Fahrenheit (3 billion degrees C) are reached, reactions are taking place very rapidly and in the fantastically intense heat many of the newly formed atoms break down again to smaller atoms. Under these conditions the atoms which accumulate in the largest quantities are those which are especially stable, such as iron, cobalt, and nickel. No star can maintain these fierce temperatures for very long. Eventually it explodes in an enormous flash of energy equal in brilliance to the light of an entire galaxy. The elements that were cooked in the superheated core, including the heavier atoms, are blown far away into space to join the vast clouds of dust and gas that wander throughout the universe. So the primordial hydrogen gas of the original dust cloud is slowly enriched with more and more complex atoms fired in the fierce crucibles of the exploding stars.[18]

One of the most surprising facts that have been learned very recently is that molecules – as well as the heavier atoms – are present in the galactic dust clouds. Until this discovery was made scientists believed that molecules, particularly the more complex ones, could not exist in interstellar space because as soon as they were formed they would be fragmented by X-ray and ultra-violet radiation. The chances of the proper atoms encountering each other in space and joining together to make a molecule were also considered to be impossibly small. However, convincing evidence has been found by radio telescopes that many molecules – even some of the delicate organic molecules – are present in the dark nebulae that float in the spiral arms of the Milky Way. These molecules, tumbling end over end in space, send forth radio waves. Each type of molecule has its own characteristic signal and these waves can be picked up by our big radio telescopes. Over thirty different organic compounds have now been identified in space, including ammonia, ether, wood alcohol, and formaldehyde.[19]

Faced with this unexpected information, astronomers had to make room in their theories for the creation and preservation of these molecules in space. They suggest that cosmic particles shooting through interstellar gas leave a trail of ions behind them. Ions are much more active chemically than neutral atoms. They combine with greater facility to build large molecules. After these molecules are formed they may be sufficiently shielded by dust particles so that they are not destroyed by ultra-violet light, just as molecules on earth are protected by the presence of the atmosphere. (The galactic dust clouds are ever so much thinner than our atmosphere, but they are also ever so much bigger.) Many scientists are not entirely satisfied with this explanation, however, and new theories may be necessary to explain these recently discovered facts.

We know now that the clouds of interstellar matter are impregnated with heavy atoms such as iron and lead and with delicate complex molecules like ether and alcohol. From these enriched clouds of galactic matter more stars are formed. Like raindrops condensing out of clouds, these 'second generation' stars are born. Although the precise mechanism is a subject of debate, the general principles seem clear. Gravity acts on swirls of matter, drawing the particles closer together. At some point, perhaps, the condensation is triggered by the presence of nuclei of denser matter just as raindrops condense out of a terrestrial cloud that is seeded with dust or ice crystals. There is evidence from the infra-red telescopes that hot new proto-stars are forming right now in the dark interior of the Great Nebula that lies in the sword of Orion. And the brilliant Pleiades are believed to be newborn stars still immersed in the faint luminous remainder of the gas cloud from which they were formed. But the film of cosmic dust which seems to pervade all the space in the universe is not 'just matter left over' from the creation of the stars any more than a mother is just matter left over from the birth of her child. Space with its tenuous membrane of matter carries the embryo of inconceivable futures within its dark womb.

In the nebula that gave rise to the sun and its planets – so the theory goes – much of the material flowed inwards to form the sun, making a concentration of matter so large that it retained even the lightest element, hydrogen, within its gravitational field. Some of the material remained in orbit, the temperature of this solar nebula decreasing with distance from the sun. Gradually the planets began to take shape by accretion of the dust particles in this cloud, the composition of each planet depending on the materials which could condense at that location in the cloud. Rock-ice mixtures formed the central core of the outer planets, which were large enough to hold the light elements also and incorporate them into the mass of the planet. The inner planets formed in a region near the sun where the intense heat helped to drive the lighter elements away. The planetary bodies themselves (for some reason that has not been resolved) stopped growing before they reached large proportions, and their gravitational fields were small enough to allow any light elements present in the original mass to float away into space, leaving the heavier dust particles to form these small dense planets. Mercury and Venus took shape in the hottest portion of the solar cloud, and water was probably not able to condense out at the high temperatures prevailing there. But the earth happened to form in an especially favoured location, not too hot or too cold. Here water could exist in a liquid state.[20]

Throughout the first few million years our planet passed through several phases of rearrangement of the initial materials. Heat generated by the condensation of the planet and by radioactive elements within its mass caused the heavier elements like iron to melt. They flowed inward towards the centre and formed the metallic core. The lighter silicates melted and floated to the top, forming the crust. The heated rocks gradually lost some of the gaseous elements that had clung to their surfaces. Substances like water, carbon dioxide, ammonia, methane, and sulphur dioxide were released from chemical combination with the constituents of the rocks. Gases and liquids forced their way up through fissures in the rocky crust, spurting forth in hot springs, geysers, smoking fumaroles, and volcanic

33. *The El Tatio geyser field, North Chile. The plumes of steam shown are capable of driving a fifteen megawatt power station, enough for a respectable town.*

eruptions. Even today plumes of gases and water are released from below the earth's surface. Geysers like Old Faithful shoot pillars of steam a hundred feet into the air. Mount Etna, Kilauea, and La Soufrière spew forth hot sulphurous gases. Volcanic plumes are approximately 90 per cent water. The remaining 10 per cent is composed of carbon dioxide, ammonia, sulphur compounds, and traces of other substances. Much of the water that is thus ejected from the crust of the planet today has been recirculated. It has previously existed in the form of cloud droplets and rain. Then it has percolated down through the ground, where it has collected in pockets and reservoirs. Eventually it is ejected again into the air. But when the planet was young a much larger percentage of this flow was 'juvenile' water, removed for the first time from combination with the rocks and soil. Volcanic activity was very intense. Gradually an atmosphere and surface water accumulated and evolved.[21]

A similar out-gassing occurred on the other planets. There is evidence of extensive volcanism and crustal deformation on Venus, Mars, and Mercury in photographs taken on the space missions. Mars has a volcanic cone that dwarfs any volcano seen on earth, some 350 miles in diameter and four times as high as Mount Everest. In fact, all the atmospheres of the inner planets are believed to have had a volcanic origin.[22]

It is hard to imagine that the cool blue waters that lap our shores and the soft zephyrs that freshen a summer night were born in the reeking conflagrations of volcanic eruptions. Volcanism seems like the very incarnation of an evil force, a voracious beast lying in wait to suddenly spring up and devour whole towns – to spread viscous flows of black lava like excrement on the fragile abodes of men, on green pastures and flowering hills. Yet this obscene force was the prime creative factor in the generation of the gases that surround our planet.

A few people have had the experience of seeing the earth open up right at their feet and have watched the violent formation of a new piece of the earth's crust. They have felt the hot breath of the planet belching forth a huge volume of gas and water to join those that already move in restless patterns around the globe. Volcanologists Maurice and Katia Krafft describe such an experience in 1973 on the island of Heimaey in the Vestmann Archipelago south of Iceland:

Between midnight and 1:35 A.M. on January 23, the earth shakes three times on the island of Heimaey. Being so used to this kind of thing, the inhabitants scarcely notice it and go on sleeping peacefully. But two men, two fishermen are walking in the streets in the eastern part of the town of Vestmannaeyjar. At 1:55 A.M., they suddenly see a line of fire behind the houses; it spreads, noiselessly and rapidly, from south to north. They have been drinking a little, of course. Convinced they are having hallucinations, they go closer to the phenomenon, and then they understand that the line of fire is a gaping fissure that spits a curtain of molten lava more than 100 meters high. A new volcano is being formed! The two men forget to give the alarm; they run home, wake their wives and children, then rush to the harbor. A few seconds later the telephone rings at the police station of the town; the incredible news spreads. The municipal authorities are immediately alerted; the police, blowing their sirens, rush through the streets of the sleeping town; the whole population awakens, jumps out of bed, and looks with astonishment at a growing, reddening volcano . . . The civil-defense department, the airport of Keflavik, the hospitals, and the police are mobilized without delay . . . Six hours after the beginning of the eruption, the last inhabitants are evacuated; approximately 5,000 people have been transported to the mainland [leaving 150 men on the island] . . .

On January 24, only three fountains of lava are still quite active over a length of 300 meters, in the northeast part of the fissure. Volcanic activity intensifies; streams of basalt lava are spewed forth by the fissure, flow east, north, and northwest, and reach the sea. Their contact with the cold water of the ocean provokes an imposing release of steam, which merges with volcanic gases to produce a plume rising to a height of 8 kilometers . . .

On the twenty-sixth, the fall of ashes reaches an accumulation of 1 meter an hour near the volcano. Windows are broken by volcanic bombs and by the shock wave of explosions; ten new houses succumb to the flames. The ground water in the pits mounts 10 meters, and its temperature rises $10°C$. The chaotic stream of lava has now advanced 500 meters into the sea. The next day the wind blows from the west and carries the ashes toward the ocean, away from the town; everyone hopes this will last, but twenty-four hours later volcanic dust again falls on the town. The volcano has already emitted fifteen million cubic meters of lava; volcanic bombs fall as far as the center of the town. On the twenty-eighth the authorities estimate that sixty-two houses have been destroyed, most of them crushed by ashes, some twenty of them burned . . . On January 30 the summit of the volcanic cone is 185 meters high, and nearly 110 houses have already disappeared under two million cubic meters of ashes. The thick stream of lava advances inexorably in the ocean, thus enlarging the island by more than a square kilometer . . .

On March 25 and 26 a lava lake rises in the crater, then overflows. A stream of lava forms and flows to the northwest, straight for the town . . . The situation becomes dramatic when the flow threatens to fill the bay of the port, thus dooming forever the economy of Vestmannaeyjar, which is based entirely on fishing. Iceland then appeals to the United States. It asks for and obtains forty-seven powerful pumps, which are brought by cargo plane and immediately installed on the wharves. These pumps have a total output of 4,500 tons of water an hour. Three large pipes, 30 centimeters in diameter and several hundred meters long, are used to convey seawater from the harbor to the terrible lava flow, and the incredible feat begins. Every hour 4,500 tons of seawater are discharged against the lava front, now advancing at a rate of 3 to 8 meters a day; this has the effect of lowering the temperature of the molten rock from 1,000° to 800°C. A volume of 20,000 cubic meters of molten rock is thus cooled by 200°C and accordingly solidified every hour ... The operation is a success, since behind the wall of solidified lava, nearly 40 meters high, the molten rock is arrested and diverted. The port is saved. This is the first time in the history of humanity that man has been able to master a small part of the destructive force of a volcano.

34. A classic photograph, taken at the Mont Pelée eruption of 16 December 1902: it shows a nuée ardente, *a turbulent mass of superheated gases and incandescent solid particles which can travel, almost silently, at the speed of an avalanche.*

The balance sheet, however, is heavy: three hundred houses have disappeared under the lava, more than a hundred are buried under ashes or burned . . . In all the low parts of the town, in the cellars, there is more than 90 percent of carbon dioxide, 0.1 percent of carbon monoxide, hydrogen sulfide, and methane. Volunteers accordingly install bellows in the basements of the houses they occupy. Nonetheless one man dies of asphyxiation from the toxic gases.

Several weeks later the explosive activity of the crater becomes intermittent, the flow of lava congeals. The island sinks slightly, since a great void now exists underground – where all the magma had been that has emerged from the erupting fissure; the walls of buildings crack. On July 3, 1973, Professor Thorbjorn Sigurgeirsson, who has directed the operations for arresting the lava flows, declares the eruption finished. The volcanic activity has lasted six months.

Ten days later an official name is given to the new volcano; it is to be called Eldfell, 'the mountain of fire'.[23]

Throughout the long formative aeons of the earth's history an untold number of violent events like this must have split the planet's crust and poured forth streams of gases to join the primordial atmosphere. Clouds formed, rain fell, and water accumulated in the lakes and seas.

The existence of water in the liquid state was probably the most important single factor in the development of our planet. As we know, water is a most remarkable substance, acting as a tempering influence on the climate, a solvent and carrier for many chemicals, a cradle for life. Water vapour in the air helps to reduce the radiation of solar heat back into space. If the earth had been a little closer to the sun, water vapour and carbon dioxide in the air might have resulted in a runaway greenhouse effect similar to that which occurred on Venus. But on earth the warming effect was just enough to be beneficial. In the lakes and shallow seas of the primitive earth the first living molecules began to replicate. The dimly lighted water screened them from the destructive ultra-violet light of the sun so they could multiply and evolve. Soon the presence of life began to alter the composition of the atmosphere. Through photosynthesis carbon dioxide was removed from the air and free oxygen was added to it. The oxygen reacted with the ammonia molecules in the air, forming water and leaving the nitrogen atoms of the ammonia molecule free to combine in pairs to make gaseous nitrogen that remained in the atmosphere. Oxygen also combined with methane to make carbon dioxide and more water. With the gradual accumulation of oxygen in the air, the ozone layer began to form, protecting the new film of living matter. Life was able to leave the protective medium of the seas, to invade and populate the land. So each change led inevitably to the next one until the planet earth as we now know it had evolved – the most inviting object that can be seen in space, clothed with green forests and deep indigo oceans, dappled by white cloud masses and framed in its translucent bubble of blue atmosphere. 'Like a million-faceted gem,' exclaimed Gherman Titov, 'an extraordinary array of vivid hues that were strangely gentle in their play across the receding surface of the world.'[24]

The modern picture of the cosmos is very different from the small intimate world of Bishop Ussher. As this understanding of the universe has gradually taken shape throughout the last century, many writers and philosophers have been deeply depressed by the increasing vastness of the time and space dimensions that dwarf human existence and make the great miracle of life itself appear as evanescent and insignificant as a miniature rainbow formed for an instant in a drop of morning dew.

> . . . it seems incredible, [said Sir James Jeans],
> that the universe can have been designed primarily to produce life like our own; had it been so, surely we might have expected to find a better proportion between the magnitude of the mechanism and the amount of the product. At first glance at least, life seems to be an utterly unimportant by-product.[25]

Scientific thought, however, has progressed dramatically since James Jeans wrote and certain aspects of the story have changed. New discoveries suggest that creation is a process which is taking place throughout time, not just a single event which occurred at one moment in the past with all history being simply the spinning out of the inevitable consequences of that one event. We know now that stars go through at least two generations. Cosmic dust clouds are changing as time goes on, becoming gradually more and more enriched with complex matter. Organic molecules seem to be forming in space, and life may be evolving on many other planets throughout the universe. The stuff of which our earth and our bodies are composed has gone through a continuous evolutionary process that has built more highly structured kinds of matter over 18 or 20 billion years. It has progressed step by step from the disorganized plasma spewed out in the Big Bang to the hydrogen atoms that made the first stars. The more complex atoms fired in these stars and scattered into space were swept up and concentrated in the planets. On earth and perhaps on some other planets in the universe, the degree of complexity of matter has continuously evolved from that time to the present, leading from elementary substances like hydrogen and oxygen to simple compounds like water, which has three atoms in a molecule, to complex organic molecules containing millions of atoms, to the self-replicating molecule – life. Then the first 'simple' cells, which actually are not simple at all but represent the organized association of at least a thousand billion atoms. And still the building process went on through the various stages of evolution to the human body – finally to the human mind, which is possibly the most highly organized state of matter in the universe. This amazingly creative form of matter is now reaching out into the cosmos, sending its messengers from planet to planet, its messages far into space beyond the Pleiades, beyond the red giants Betelgeuse in Orion and Antares in Scorpius near the centre of the Galaxy. These little waves of human thought interlace with the cryptic signals sent back by the quasars as they race outwards towards the edge of the universe and with the faint voices of the newborn molecules that tumble in their lonely paths through the vast twilight spaces between the stars.

Whatever else is still unknown about creation – and there is a great deal – one principle does shine through. Creation is a continuing, ongoing process. New stars are being born today in the dark clouds that veil the spiral arms of the Milky Way, and thoughts are being born in the minds of men – thoughts that can shake the planet – perhaps even the universe itself – and alter the course of its history.

CHAPTER FIFTEEN

FIRE OR ICE

Observe always that everything is the result of a change, and get used to thinking that there is nothing Nature loves so well as to change existing forms and to make new ones like them.

– Marcus Aurelius[1]

In the tiny country of Bangladesh, sandwiched between the sprawling delta of the Ganges River and the towering range of the Himalayas, rainfall every year is extraordinarily high – 200 inches is not unusual and the culture of rice is planned to take advantage of the fields that are flooded every summer. But even with a good rice harvest there is never enough to feed the 74 million people that crowd thick as flies in this small corner of the earth's surface. Annually before the new harvest comes in, there is a long period known as 'the hunger months' when the comparatively well-to-do tighten their belts and eat just one meal a day; the very poor, deprived of even that one meal, do not survive until the harvest is in. Every year famine is a fact of life in Bangladesh, but in 1974 a late-summer flood inundated almost half the country. The ripening crop standing in the fields was totally destroyed and most of the inadequate stores of grain was lost.

Steve Raymer, a photographer for *National Geographic* who visited the stricken area, told of seeing thousands of starving people lying on the streets and the bare ground surrounded by their own faeces and plagued with flies they were too weak to brush away. There were naked children whose withered buttocks and wrinkled legs looked like very old men's and infants so thin their chests resembled bird-cages beneath their festering skin.

In Dacca, the capital, he watched train-loads of refugees pouring in from the even more desperate country towns. The trains were overflowing, dripping with people. They clung to the window frames and found toeholds on the outside coping; they were packed in a tight mass on the roofs of the cars. When they left the train at the Dacca station, many of these refugees went no farther. They just squatted down on the railroad platform and took up their abode there. Amid the thousands of defeated, emaciated people, Raymer related, a woman lay on the station platform and gave birth. Several other women came to her aid. One found a rusty sickle near the tracks and cut the umbilical cord. Another, seeing the mother was too weak and shrivelled to nurse, dabbed the infant's lips with coarse table sugar. Burning with fever, the helpless mother watched . . . Bangladesh had another mouth to feed.[2]

Starvation is a long-drawn-out horror, extending over many weeks. Kamala Markandaya, the India-born novelist who saw poor people caught in the grip of hunger throughout her native land, describes what it is like:

> For hunger is a curious thing: at first it is with you all the time, waking and sleeping and in your dreams, and your belly cries out insistently, and there is a gnawing and a pain as if your very vitals were being devoured, and you must stop it at any cost, and you buy a moment's respite even while you know and fear the sequel. Then the pain is no longer sharp but dull, and this too is with you always, so that you think of food many times a day and each time a terrible sickness assails you, and because you know this you try to avoid the thought, but you cannot, it is with you. Then that too is gone, all pain, all desire, only a great emptiness is left, like the sky, like a well in drought, and it is now that the strength drains from your limbs, and you try to rise and find you cannot, or to swallow water and your throat is powerless, and both the swallow and the effort of retaining the liquid tax you to the uttermost.[3]

Five thousand miles south-west of Bangladesh in the Sahel region of Africa, the cumulative effects of six years of drought claimed millions of lives in 1973 and 1974. Many of these were children, because when famine strikes a land the children are the first to suffer.

William Mullen, a reporter for the *Chicago Tribune*, visited a refugee camp in Timbuktu, the desert crossroads whose name has long been a synonym for the place at the end of the world. There he found the destitute remnant of the once proud tribe of the Tuaregs:

The drought appears to have ended forever the romantic epoch of nomadic tribesmen who for centuries crossed the Sahara on camel caravans and roamed the desert fringes with tents and herds of cattle.... The Tuareg nomads were the noblemen of the Sahara.... Their camel caravans carried the goods of Asia and Europe to fabled Sahara cities such as Timbuktu, Gao, and Agadez. Now the Tuareg rules neither the desert nor his own destiny. The desert itself, growing at a rate of 30 miles a year, has killed off the herds of cattle he counted as his wealth and choked off the supplies of water he needed to survive.... They allowed their burgeoning herds to over-graze the fragile pastures. The drought came to dry up the grassland, and the Tuaregs cut down all the savannah trees and bushes so their starving livestock could eat the leaves. As the drought persisted, they lost not only the animals, but also the savannah because of their careless treatment. With nothing to protect the topsoil, it blew away with the desert winds that brought down the sand dunes.... As the grasslands disappeared, their livestock disappeared, they had nothing left to trade, and their desert camel caravans carrying salt, silk, and spices went the way of oblivion.[4]

Mullen saw thousands from this once fierce, indomitable culture crowded into refugee camps living in filth and gnawing on bones, waiting for death. Too weak to stand, women and children lay close together on the dusty ground. 'There was no comfort left but a gentle touch and a loving presence.'

The long drought in the Sahel came to an end in 1975 with a series of torrential rains which beat down upon the desiccated, unprotected land, producing raging rivers overnight and cutting the few roads that were serving as a lifeline for relief supplies. Because of the rains many more people were destined to starve. That year the government in Mali forced the Tuareg tribes to leave the refugee camps and return to the desert, where they spent their days searching the barren countryside for anthills to open up so they could take the stores of grain which the insects had packed inside. Soon the supply of anthills would be used up, Mullen reported, and the Tuaregs 'would have to resort to collecting nettles, sort of oversized sandbur-like weeds, as their only source of food. Surviving on nettles would be difficult ... because it takes a strong person two days of grinding to transform enough of them into flour for one meal.'

Flood and drought – always disastrous – have struck with a heavy hand all around the world in recent years. Beginning in 1968, failure of the monsoons occurred for six years in the semi-arid lands south of the Sahara. In 1972 a change in the ocean currents off the coast of Peru caused a serious failure of the anchovy catch, which normally represents a substantial portion of the world's fishing industry. Russia and other Eastern European countries experienced a shortage of winter snow cover, which caused fatal exposure of young winter wheat seedlings to freezing weather. A massive crop failure occurred, necessitating large purchases of grain. After one reasonably normal year (1973), in 1974 the climate was erratic in most of North America – excessive spring rains, drought in July, and early fall frosts. The next year brought drought conditions to East Germany, Yugoslavia, Hungary, and the midwestern United States. Satellite pictures taken of China showed that many of the agricultural areas of this most populous nation were burned out that year. Rains fell too heavily in some parts of Russia and not enough in others. The year 1975 brought average growing conditions to North America and Asia, but variable and poor conditions prevailed in Europe, Latin America, parts of Africa, the Near East, and Australia. In the Soviet Union hot dry weather cut grain production by 10 per cent.[5] In 1976 a prolonged winter drought killed much of the winter wheat crop on the Great Plains of the United States. In the spring and summer drought struck California, South Dakota, Minnesota, and Wisconsin, where eighteen counties were declared disaster areas. Western Europe endured the most prolonged spell of hot dry weather that had been experienced in living memory. Animals were slaughtered for lack of pasture; billions of dollars in wheat, potato, and sugar beet crops were lost; water was rationed; and widespread fires raged uncontrolled through forests in France and Britain. On the other side of the earth, Australia, too, was scorched by a disastrous drought. In California the dry weather continued through the winter months, which usually provide a large proportion of the annual rainfall. While farther east, most of the United States suffered through the most severe winter of the century.[6]

The weather seems to be going sour all over the world. Surely something strange and unprecedented must be causing these deviations from 'normal weather'. But climatologists who made a study of this problem arrived at quite a different conclusion. It was the favourable and relatively uniform climate between 1955 and 1970 that had been unprecedented.[7] During these years the temperature and rainfall conditions had been very consistent and beneficial. A comparable stretch of weather had not occurred during recorded history.

Throughout this same good-weather period, technological advances were improving the yield per acre. In spite of a rapidly growing population the production of food per capita was keeping pace throughout the 1950s and even well into the 1960s. The Green Revolution had achieved the miracle of feeding India. By utilizing special hybrid strains of wheat and rice, crop yields were doubled and tripled. The productivity of these strains depends very critically on moisture and temperature conditions as well as on the application of generous amounts of fertilizer. During the first years of the experiment these conditions were met.

Since scientific advances and good weather happened to occur simultaneously, the mistaken assumption was made that technology had been able to reduce the *variation* in crop yield from year to year as well as to improve the overall productivity. As Stephen Schneider points out in his book *The Genesis Strategy*, this assumption led to an unjustified optimism about the ability of human societies to feed indefinitely their growing populations. The principal grain-producing nations did not make any effort to build up extra reserves against a rainy day or a day when the rains would not come. The United States did have a surplus left over from the 1940s, but when the long spell of good weather broke in the early 1970s, a large part of this reserve was used up. Now, in spite of great technological advances, more people are threatened with famine than at any other time in the earth's history.

Weather is possibly the most dominant factor in determining the quality of human existence, but although we have set foot on the moon, have identified organic molecules in the tenuous dust of interstellar space, and measured the speed of galaxies a billion light-years away, we still do not understand the intricate interplay of forces that move the earth's weather.

We can, however, chart its history in rather surprising detail. Climate information is available from recorded history for the past three or four thousand years. Chinese writings go back to the fourteenth century B.C. For even more far-off times climatologists study core samples (long cylindrical borings dug from the soil, from glaciers, and from the bottom of lakes and seas). By analysing the chemistry and the residues of biological materials present in each layer of these cores, they can deduce the climatic conditions that existed when it was laid down.[8] For example, the presence of different types of pollen – spruce, maple, ragweed, and so on – indicates temperature and humidity conditions favourable to those particular types of plants.

This research has turned up the surprising and alarming fact that dramatic climatic changes have occurred much more rapidly than had previously been supposed.[9] Within twenty years very sharp changes in temperature have occurred and the earth may have passed from an ice age climate to an interglacial type of climate like the present in just over a century. Until these studies were made, it had always been assumed that important changes took place too slowly to have a major impact on human affairs. Now climate experts like Reid Bryson, who directs the Institute for Environmental Studies at the University of Wisconsin, believe that in less than a generation alterations in weather patterns may seriously affect the number of people the earth can support. Furthermore, theories suggest that the activities of man can contribute very substantially to weather changes. It has suddenly become imperative that we understand the processes that regulate the climate.

The geophysical record shows that our planet has passed through three – perhaps four – major epochs of glaciation at very widely spaced intervals. Each epoch lasts for millions of years and consists of several glacial and interglacial periods. The epochs may have been caused by a change in the amount of radiation emitted by the sun, or perhaps, as noted earlier, they were caused by conditions in the galactic space through which the solar system passes in its orbit around the Milky Way. A third explanation is that episodes of intense volcanic activity brought on these major glaciations. Whatever the cause, these large cycles of climatic change have moved in a very slow rhythm – approximately 250 million years. Superimposed on this cosmic pulse are many smaller fluctuations, and these are the ones that have the greatest relevance to the human condition.

The last glacial epoch began about two million years ago, and since then the earth's climate has oscillated between ice ages and warmer interglacial periods. The cause of these oscillations within a major ice epoch is not definitely known, but the most favoured explanation is a regular rhythmic change in the earth's orbit and tilt of its axis in relation to the sun. This hypothesis was proposed by the Yugoslavian scientist Milutin Milankovitch in the 1920s. At first it evoked only passing interest because at that time it was believed that there had only been four recent ice ages and the timing of these did not correspond well with Milankovitch's theory.[10] But more accurate information now indicates that there have been many ice ages within the last glacial epoch and their spacing does seem to correlate remarkably well with the cyclic changes in the shape of the earth's orbit, the tilt and precession of its axis as it moves around the sun. These slight changes in the orientation of the planet result in regular and predictable variations in the amount of solar radiation reaching various portions of the earth's surface. They set up a complex rhythm of superimposed climatic cycles of 100,000, 42,000, 23,000 years respectively. According to this model, which is quite widely accepted by scientists today, the forecast for the next several thousand years is a chilling one – steadily declining temperatures and extensive glaciation in the Northern Hemisphere.

The latest ice age reached its peak 18,000 years ago. Huge sheets of ice which were often 10,000 feet thick covered much of the mid and high latitudes around the globe. About 15,000 years before the present (B.P.) the glaciers began to retreat. Again this change may have involved several episodes of retreat and advance of the ice. The most recent of these very short-lived oscillations appears to have occurred 11,600 B.P. and may have been the reason for the Great Flood which is described in the folklore and religion of many ancient peoples. Analysis of several cores taken from the Gulf of Mexico indicates a sudden surge of climatic change and a dramatic rise in sea level, possibly caused by the disintegration of one of the polar ice caps. Paradoxically, the melting of a large glacier could be caused by either a general warming or a general cooling trend. Warmer temperatures at the poles would melt the ice caps, causing them to soften, break up, and fall into the sea. A cooling trend, on the other hand, could cause such a thick build-up of ice that the bottom of the glacier would be melted by the pressure of the weight above it. Then the glacier, which normally moves from higher to lower ground very slowly by 'plastic flow', would move much more rapidly as the water at its base provided a liquid lubricant. The huge mass of ice would move more rapidly downhill and fall into the sea. In either case this large quantity of ice suddenly dumped into the ocean would raise the water level tremendously and reduce the water temperature, bringing icing conditions to lands at lower latitudes. There is evidence that in 11,600 B.P., following a general warming trend when the major ice shield had been retreating northward, a sudden surge of glaciation re-covered all of northern North America in a sheet of ice.[11]

Palaeoclimatologist Cesare Emiliani, who directed these studies, speculates that the flooding of low-lying coastal areas, many of which were inhabited by man, gave rise to the deluge stories common to many traditions. 'Plato set the age of the flood at precisely 11,600 B.P.'[12]

After this episode a temperate period began which has continued with some smaller fluctuations up to the present time. Focusing in now on the last two thousand years, historical records show that most of northern Europe enjoyed an unusually warm climate between the years 400 and 1200. During that time the northern seas were generally free of ice. Storms were rare. The Vikings set forth from their home bases in Scandinavia in open longboats and established colonies in Iceland and Greenland. As the weather conditions began to deteriorate in the thirteenth and fourteenth centuries, the bases and agricultural communities that they had founded were abandoned.

During the years 1200–1400 the climate in Europe was extremely variable. Floods alternated with droughts, mild with severe winters, and storms were much more frequent. A prolonged cold period in the early fifteenth century was followed by several decades of generally milder weather. Then in 1550 the weather began to grow colder and this trend continued throughout the next three hundred years, an epoch now known as the Little Ice Age. During this time temperatures in northern Europe and America were considerably colder than they are today.[13] The arctic ice pack was greatly extended. Paintings and engravings of that time show scenes of ice skating on the river Thames, which has not frozen over in recent times. An engraving made about 1855 depicts the Argentière glacier near Chamonix in the French Alps as considerably more extensive than it is today. Evidence collected by the French historian Emmanuel Leroy Ladurie documents the fact that Alpine glaciers grew almost continuously between 1590 and 1850.[14] In the United States reliable records indicate that average midwestern temperatures in the 1830s were as much as seven degrees colder than today.[15]

Then in the last two decades of the nineteenth century the trend was reversed and the world passed through a long warming period that continued until about 1940. The glaciers retreated; the length of the growing season in England increased by two to three weeks. Throughout the 1940s the coasts of Europe and Asia were open to shipping for an extended summer season. Ports in West Spitsbergen were free of ice for seven months of the year, compared with the three-month period at the turn of the century. Ice floes in the Russian portion of the Arctic Ocean decreased by 350 square miles between 1924 and 1944. During these years various species of birds and fish typical of warmer climates extended their range into higher latitudes.[16] The black-browed albatross, the ovenbird, and the Baltimore oriole summered in Greenland; wood warblers, skylarks, and scarlet grosbeaks visited Iceland – while the real arctic species moved farther north.

In 1912 the first codfish appeared off the coast of Greenland – a strange fish never before seen by the local inhabitants. By the 1930s cod fishing had become a major occupation and the centre of the fishery was moving steadily north.

In the early 1940s another reversal occurred, and the climate – at least in the Northern Hemisphere – turned gradually colder. (Records for the Southern Hemisphere are not sufficient to document the trend there.) The most pronounced temperature changes have occurred in the high latitudes. This variation by latitude is typical of changing climatic patterns. The greatest alterations occur towards the poles, while in equatorial regions there may be little or no change. The average temperature in Iceland, for example, was four or five degrees colder in 1973 than it was around 1940, but the Northern Hemisphere as a whole cooled less than one degree. According to Hubert Lamb, director of the Climatic Research Unit at the University of East Anglia, the growing season in England dropped back by about two weeks during that time.[17] The temperature of the western North Atlantic declined 3.6°F (2°C) between 1950 and 1970.[18]

Variations of one degree or less in mean global temperature are especially significant in marginal areas – at high latitudes, where the growing season is already a little too short, and in semi-arid lands, which even in good years receive just barely enough rainfall. As the higher latitudes become colder, the temperature difference between the polar and the equatorial regions increases and this leads to more energetic wind-patterns. The band of cyclones and anticyclones that moves on prevailing westerlies around the mid-latitudes circulates farther south and the storm patterns are more intense. The weather may increase in variability with droughts, floods, and freak freezes imperilling the crops. Along with the band of turbulent low-altitude westerlies, the position of the principal jet streams is also shifted farther south. The easterly jet, which develops in the summertime and normally flows across India and the Sahel, bringing the life-giving monsoon rains, moves nearer the equator. Some climatologists believe that this shift was the cause of drought and famine in those lands.

Such a small change – just a degree or less in world temperature – is enough to destroy the traditional way of life of millions of people. The Tuaregs lost their caravans; much of their green savanna land joined the sand dunes of the Sahara. The tall Masai, herdsmen of the East African Plains, watched 90 per cent of their cattle die during the drought and many of those tribesmen must take on a new life style in order to survive. Even in the 'advanced countries' farmers have watched their crops shrivel and die; their farms blow away on dark, dust-laden winds.

Is it possible for mankind to anticipate these variations in the earth's climate? To plan for them? Perhaps even to control them? At the present time the field of climatology is fraught with controversy. However, certain generally accepted facts and relationships have been discovered.

Almost all of the warmth experienced on earth comes directly or indirectly from the sun, and so it is logical to postulate variations in solar luminosity as the cause of fluctuating climatic conditions on earth. Indeed it would be unreasonable to expect that our star would continue year after year, eon after eon, to pour forth energy at exactly the same rate. Although attempts to correlate the large climatic changes with solar activity have been only partially successful, a few interesting correlations have been found.

For example, the Little Ice Age was approximately centred on the years when sunspot activity was at a minimum.[19] About 1615, as we have already seen, sunspot numbers declined, reaching an all-time low about 1650. For the next seventy years the sun was unusually quiescent – auroral displays were very rare, sunspots and solar coronae during eclipses were virtually nonexistent. Although the Little Ice Age started earlier than 1615 and lasted longer, some weather experts believe that a decrease in solar activity during those years was at least a contributing factor in the decline of the temperature.

Several other correlations with the sunspot cycle have also been noted. The world production of wheat was greater in 1958 (one year after a sunspot maximum) than in each of the next five years. In 1968 (a year of maximum sunspot activity) the crop was larger than in any of the next four years. On the other hand, 1954, a sunspot minimum, produced a poor crop. Although the evidence is still very tenuous, there seems to be some support for the theory that wheat yield in the Northern Hemisphere has been enhanced as the sunspot maximum approached, and depressed at times of minimum activity.[20]

The droughts that have struck the Great Plains of the United States during the past 160 years have occurred in a 20- to 22-year cycle, corresponding roughly with alternate sunspot minima. Why *alternate* minima? Scientists suggest this may be related to the fact that the magnetic field associated with the solar wind reverses each sunspot cycle; so the same solar wind conditions repeat every 22 years instead of every 11.[21]

You may also remember that the region of highest incidence of killer tornadoes has moved in a clockwise direction through the central states, completing a full turn in 45 years. Again the time for a complete rotation seems to correlate with sunspot cycles. But why every *four* cycles in this case? Nobody knows the answer to this question or, indeed, whether these pulses in weather activity are truly related to sunspot activity. The correlations could be purely coincidental.

Matching rhythms on an even smaller time scale have been discovered by scientists working with Walter Orr Roberts at the National Atmospheric and Oceanic Administration and with John Wilcox at Stanford University. The sun as viewed from the earth turns on its axis every 27 days and the solar wind which sweeps past the earth changes its magnetic orientation four times during a typical solar rotation. So approximately once a week the magnetic field flips over and the time when this shift occurs can be quite precisely identified now by measurements made in satellites.[22] Between the reversal times the intensity of the solar wind rises and falls in a regular pattern. Using old weather records, the scientists have made an estimate of the total size of the low-pressure systems occurring on any one day in the Northern Hemisphere during the winter months. The number of square miles influenced by these cyclonic systems is a rough indication of the amount of bad weather occurring at that particular time. A plot of these totals day by day showed that they rose and fell on approximately a weekly rhythm, matching very closely the change in solar wind intensity and the orientation of the magnetic field. Since the sun rotates on a 27-day instead of a 28-day cycle, the bad-weather maxima tend to shift their position very slowly in relation to the week. Perhaps this is part of the explanation for those months at a stretch when it seems to rain every week-end!

These theories are very new and controversial. Many climate experts do not believe that the solar wind could affect the weather systems in the troposphere. They point out that the amount of energy in the solar wind is only one ten thousandth of the energy in the lower atmosphere. But other scientists who have seen the evidence of these matching patterns believe that a triggering mechanism may be involved. A small change in the solar wind might start a series of reactions which would magnify the effect by the time it reaches the troposphere.

Is it possible that the biblical notion of seven fat years and seven lean years represented the accumulated experience of farmers several millennia before the birth of Christ? Seven and seven, of course, does not make eleven. But – who knows? – maybe the sun's pulse ran slower three thousand years ago.

Even if the sun always poured forth its energy at an absolutely even rate the amount received at the earth's surface would be significantly altered by cloud formations and veils of dust in the atmosphere. It is an established fact that major volcanic eruptions inject a large quantity of aerosols and particulates high into the stratosphere where they reflect the sunlight, causing beautiful purple sunsets, and clouds above the troposphere, and cooler weather for a year or so over the whole world. The eruption of Krakatoa in the Sunda Strait near Java in 1883 (one of the largest volcanic eruptions in recent history) was followed for several years by poor growing seasons in many lands. A sharp freeze occurred in California in 1884, as evidenced by studies of tree rings. In 1963 after the eruption of Mount Agung on Bali, observations at Mauna Loa Observatory in Hawaii showed a decrease of about half a per cent in the solar radiation reaching the troposphere.[23]

From the end of the nineteenth century until 1963 there was very little volcanic activity on earth. This lull may have been one of the factors contributing to the warming trend from 1885 to 1940. In fact, some climatologists believe that episodes of strong volcanic activity generated and helped to sustain the great ice ages. Studies of glacial cores show that at several times in the earth's history a high level of volcanic activity occurred simultaneously with minimum temperatures, and the amount of volcanism associated with the Quaternary (the time of the last ice ages) was much greater than it had been during the previous 20 million years.[24]

Theoretically, a cooling trend, once started, can feed on itself and – if all other factors are constant – lead to a progressive decline in the world's temperatures. During cold years the amount of snow cover increases, extending to lower latitudes and melting later in the spring. The snow and ice-covered ground reflect the solar radiation much more efficiently than bare ground. The oceans and the air that blows across the glaciers are cooled; so the average surface temperatures decline even more. Slightly larger snow caps remain throughout the summer. The next year the same process continues and the ice pack grows.

Photographs taken from satellites over the past decade showed that the snow cover in the Northern Hemisphere was beginning to build up earlier in the fall and melt later in the spring from 1971 to 1974. A very sudden increase, almost 12 per cent, in the extent of the white areas occurred in 1971. For the next two years it remained approximately constant until 1974 and 1975, when there was a slight decrease.[25] It is still too soon to tell whether this latest aberration represents a true reversal in the trend or whether it is a small fluctuation in an overall temperature decline.

Volcanic action is not the only source of a dust veil between the earth and the sun. Industrial pollution adds enormous amounts of aerosols and particulates which ride the winds and prevent some of the solar radiation from reaching the ground. The presence of these pollutants will probably become more widespread, in spite of environmental regulations, because of rapid industrial growth and expanding populations.

Even in the non-industrial portions of the world, man's activities are contributing to the thickening pall of dust. Slash-and-burn agricultural practices are commonly employed in primitive farming areas, burdening the air with vast quantities of smoke and ash. Under the pressure of

35. Industrial pollution.

increasing population, semi-arid lands are being over-grazed. As in Mali, Niger, and Ethiopia, marginal lands are used so intensively that a slight change in precipitation patterns can turn them into bare stretches of dust and sand. According to a study made by the United Nations, the deserts are growing all over the world; 6.7 per cent of the earth's surface may be man-made desert.[26] In the last half century 450,000 square miles have been added to the Sahara and the loss of productive capacity by the semi-arid lands adjacent to the deserts is an even more serious problem. Population pressure and drought on the fringe of the Thar Desert in India is destroying 30,000 cultivable acres a year.[27]

The loose unprotected soil and sand created by these processes is gathered up by winds and distributed throughout the troposphere, blocking out the sunlight. For weeks at a time the sky is sand-coloured instead of blue and the worst dust storms bring darkness at noon. Records compiled by the National Oceanic and Atmospheric Administration show that sunlight decreased 1.3 per cent in the United States between 1964 and 1970.[28]

From his studies of lake-bottom cores Reid Bryson believes that the presence of the dust veil caused by human activity may be sufficient to trigger the onset of a new glacial period. Man-made pollution, he says, is comparable to moderate volcanic activity. Each of these sources loads the atmosphere with somewhere between 8 and 24 million tons of particulates. Bryson concludes that this dust cloud can account for a large part of the one-degree decline in average world temperatures that took place between 1940 and 1973, and if this cooling continues, in just two or three decades the earth's temperatures may reach levels similar to those which prevailed in the Little Ice Age. The growing season would be sharply reduced in the mid and high latitudes. In countries like Canada and Russia crops now successfully raised could no longer be grown. The jet streams and the monsoons would be displaced much farther south, bringing worse droughts to India and the Sahel. As many as a half billion people could starve by the end of the century.[29]

If the discharge of man-made dust into the air were to continue without any compensating factors for a hundred years or more, the world could be plunged into a major ice age. Glaciers would begin to form over the mid and high latitudes around the globe; all the mountain ranges would be covered with perpetual snow. Inexorably, year by year and century by century, the ice pack would grow. Judging from the conditions that prevailed in the last major glaciation, New York City, Berlin, Edinburgh, and all the land north of them would eventually be buried beneath ice several hundred yards thick. So much water would be bound up in the glaciers that ocean levels would drop, baring lake bottoms and continental shelves. Less water in the oceans would mean less evaporation and decreased rainfall. The lands that were free of ice would be subjected to prolonged drought. The deserts would grow, the forests shrink, and much of the land would consist of semi-arid loess – a yellowish brown soil dotted with scrubby tufts of grass. Unimpeded winds would blow with relentless fury across the barren ice fields and the desiccated land. In the mid and high latitudes winters would be long, summers short, and the differences between day and night temperatures would be greater than we have today.

The planet itself would survive this ordeal by ice just as it has a dozen other times in the past two million years, but its delicate cargo of living things would be cut to just a fraction of its present size. The blueberries that grow on the rocky hillsides of Maine, the skylarks that winter in Iceland, the white dogwood that illuminates the northern woods in May – none of these would survive. The children who now run barefoot across fields of grass in spring and wade in waters warmed by the summer sun – only a few of these would be able to find enough food and shelter as they grew older in an increasingly forbidding world.

This is the death by ice (the snow blitz) forecast by some climate experts, but there are others who do not agree with the prediction. Insufficient weight has been given, they say, to several factors that work in exactly the opposite direction.

For example, the presence of dust clouds in the atmosphere may under some circumstances warm the planet's surface instead of cooling it. There is a complex relationship between the reflectivity of the cloud itself and the reflectivity of the ground surface which is shadowed by the cloud. If the ground reflects more of the solar energy than the cloud does, then the presence of the cloud may actually add warmth to the earth system. Ice and snow reflect almost 80 per cent of the incident radiation, whereas oceans reflect only 5 to 10 per cent. Clouds also vary considerably in the way they turn back, absorb, or transmit the solar energy. Aerosols containing sulphur compounds are more reflective than water vapour. All of these variables as well as the height and density of the clouds must be taken into consideration in calculating their total impact on the earth's weather.

A striking illustration of the effect of different types of cloud formations has been revealed by the exploration of the other planets. The dense dust clouds encountered by Mariner 9 on Mars cooled the surface of that planet, while the clouds that completely shroud Venus all of the time produce a greenhouse effect which causes the fiercely hot ground-level temperatures. As we have already seen, the presence of carbon dioxide, water vapour, and certain other chemicals in the air tend to make it act like a one-way filter, allowing solar energy to pass through and be received at the surface but preventing the reflected energy from escaping into space. Some climatologists believe that this greenhouse effect will become the dominant factor determining the earth's climate over the next half century.

The burning of fossil fuels is continuously adding carbon dioxide to the air, and it is known that amounts have been building up quite steeply since the middle of the nineteenth century, when the industrial age began. In 1860 the carbon dioxide concentration in the atmosphere was about 292 parts per million by volume. By 1950 it had risen to 310 parts per million, and by the end of this century it is expected to reach approximately 400. Estimates indicate that this amount of carbon dioxide would induce a global temperature rise of about 1.8°F (1°C). By the year 3000, at the present accelerating rates of fuel consumption, the concentration of carbon dioxide may have increased to 1,850 parts per million, an amount sufficient to raise world temperatures 16°F (9°C).[30]

On this subject, too, there is disagreement among the scientists. Some say the fossil-fuel supply would have been used up long before such levels were attained. Others maintain that the discovery of new deposits as well as improved recovery techniques would make this amount of fossil fuel available.

Approximately half of the carbon dioxide that is introduced into the air by combustion is dissolved in the oceans or utilized by vegetation. Laboratory experiments have shown that when the carbon dioxide content of the air is enriched, plants grow more rapidly. So some of the carbon dioxide which is continuously being added to the air by combustion is probably being utilized in this beneficial way to promote faster growth.

However, as the temperature of the oceans increases they are not able to hold as much carbon dioxide in solution. Released from the oceans, it enters the atmosphere and adds to the growing concentration there. The warmer oceans also cause more evaporation, and water vapour enhances the greenhouse effect.

To some extent this portion of the feedback process may be balanced by an opposing factor. The increased water vapour content would increase cloud formations, which reflect some of the sun's energy. It is a difficult problem to estimate which of these results would be dominant as earth temperatures increase.

One of the interesting new sidelights of the fluorocarbon question is the discovery that these compounds also make a significant contribution to the greenhouse effect. It has been estimated that if the release of fluorocarbons to the atmosphere returned to the growth pattern of the 1960s and early 1970s, by the end of the century mean global surface temperatures would increase by about 1.3°F (.7°C) – an effect comparable with that anticipated from the burning of fossil fuels.[31]

As the climate begins to warm up, even just a small amount, the glacial ice caps begin to recede. When this happens the reflectivity of the planet decreases and more solar energy is absorbed. The ice melts a little more. This positive feedback process is just the reverse of the one that could lead to an ice age.

Finally, the burning of fuel on earth puts heat directly into the earth's ground, water, and air. Factories, furnaces, power plants, automobiles, and so on all pour 'waste heat' into the earth system. Not only fossil-fuel combustion but also nuclear-power sources contribute to the earth's heat by converting mass to energy. (Hydroelectric, solar, and wind power are energy sources that do not add an extra burden of heat because they utilize energy that is already present.)

At the current rate of conversion this release of waste heat is too small to be significant on a worldwide basis, although, as we have seen, it does affect local climates, creating heat islands around the big cities. But energy usage is growing at an exponential rate, and if it continues to accelerate at the rate it has in the last few decades, then in one century the direct injection of heat by human beings into the earth system will be approximately 1 per cent of the amount received on earth from the sun. This would be enough to make a devastating difference in the conditions on this little planet.[32]

Several of the warming processes could work together, speeding the advent of a hothouse climate on earth. The lower latitudes would become dense, dripping jungles. The polar ice would begin to melt, inundating the heavily populated coastal regions of the world and the farmlands along the wide river deltas. The streets of New York, London, and Buenos Aires would be flooded and sickness would spread in the warm foetid waters. Then as the oceans rose higher and storms periodically drove great tidal floods over the seawalls, the coastal cities and their satellite suburbs would become totally uninhabitable. They would be abandoned along with the rich alluvial plains of the Mississippi, the rice islands of Bangladesh, and the green fields of Ireland. In the low countries of Holland and Belgium, dikes could no longer hold back the sea. Hundreds of thousands of tulip bulbs which now make the fields of these lands glow with colour in springtime would lie rotting on the ocean floor. The warm sea waters, unable to hold their generous content of carbon and oxygen, would not be able to support the present variegated population of aquatic life. They would become clogged with dead and dying creatures and a terrible stench would rise from the seas. In the heavy sultry air the effluents produced by the burning of fossil fuels would hang like a pall across the land. The sulphur and nitrogen oxide fumes, reacting with the abundant water vapour, would turn into corrosive acids. Dark skies laden with clouds would discharge deluges of acid rain. Earth would begin to look and feel a little like Venus. As Carl Sagan says: 'Venus may be a cautionary tale for our technical civilization, which has the capability to profoundly alter the environment of our small planet.'[33]

These stories of ice death and heat death placed back to back may seem more like science fiction fantasies than serious projections by responsible climate experts. After reading these conflicting views, it is tempting to draw the comforting conclusion that the opposing weather trends will average each other out or, at the worst, will produce a reasonably temperate climate lying somewhere in between the two extremes. But to take this complacent attitude would be to miss the really significant new insights on which both of these predictions are based.

The earth's climate has passed through very abrupt transitions between glacial and interglacial periods, changes rapid enough to be important in a human lifetime. Apparently the climate is much more sensitive to small forces than has previously been imagined. Furthermore self-augmenting processes can convert a little ripple of temperature variation into a wave that gathers size and momentum for many decades or centuries until some major countervailing force appears to reverse it.

Although some cancelling of opposing tendencies does occur, true balance cannot be expected to last long. For example, some climatologists believe that increasing concentration of carbon dioxide in the earth's atmosphere may have counteracted the effect of the man-made dust cloud during the 1920s and 1930s. Perhaps the dust cloud was the larger of these two factors and was growing at a slightly faster rate, but the reduced volcanic activity during those years helped tip the scale in favour of the warming trend. Temperatures rose slightly during those years. Then beginning about 1940 the influence of the dust cloud began to override the greenhouse effect. Temperatures started to cool. In 1963 volcanic activity became a significant factor again and added to the cooling trend. Now, unless some counterforce intervenes, this decline in world temperatures could very quickly begin to accelerate by the feedback mechanisms. Just as on a teeter-totter, when the equilibrium point is definitely overcome the system passes through a wide swing before equilibrium is possible again.

Severe weather changes such as those predicted by the various schools of climate experts would have important international implications. The world's reserves of grain, which provided an emergency supply of 69 days in 1972, had dropped to about twenty-nine days' supply by 1974. By 1976 they had increased only slightly to thirty-nine days. The principal grain-producing nations have not taken action to build up a more adequate margin, and since weather variability can cause a 10 per cent reduction in productivity, this small reserve could easily be wiped out in one year.

In the meantime, the world is coming to rely more and more on one source of supply. As recently as the 1930s all but one of the continents exported grain (western Europe was the sole net importer). Now only North America and Australia are net exporters, and North America accounts for 88 per cent of the total. As Lester Brown points out in his book *By Bread Alone*, North America today controls a larger share of the world's exportable surplus of grains than the Middle East does of current world oil exports.[34]

The political and economic power conveyed by this concentration of a vital resource can hardly be overestimated. A report published by the Central Intelligence Agency of the United States in 1974 contained the following comments:

> In a cooler and therefore hungrier world, the US' near-monopoly position as food exporters . . . could give the US a measure of power it

had never had before – possibly an economic and political dominance greater than that of the immediate post-World War II years. In bad years when the US could not meet the demand for food of most would-be importers, Washington would acquire virtual life and death power over the fate of multitudes of the needy. Without indulging in blackmail in any sense, the US would gain extraordinary political and economic influence. For not only the poor L D C's [Low Development Countries] but also the major powers would be at least partially dependent on food imports from the US.[35]

As with all sources of political power, there is a great temptation to use the food monopoly for self-serving ends. Those who stand to profit from a food shortage can use this power passively by failing to do everything possible to build up grain reserves during favourable years. Or they can wield it aggressively by deliberately altering some of the forces that regulate the weather. In fact, manipulation of climate has already been used as an offensive weapon. During the war in Vietnam the United States used cloud seeding to increase the rainfall and turn the Ho Chi Minh Trail into a river of mud. It has also been rumoured that the C I A with cooperation from the Pentagon carried out a secret weather offensive against Cuba by seeding clouds headed towards the island on the prevailing trade winds so that the rain would precipitate out over the sea and the island would suffer a drought, imperilling the sugar harvest.[36] The harvest did fall short of the goal set by Fidel Castro that year (1970) and he offered to resign. But the science of meteorology is still too primitive to prove that the cloud seeding was responsible for the reduced yield.

Even more drastic schemes for tampering with the weather have been seriously proposed – for example, melting the north polar glaciers by spreading them with ten million tons of black soot. It has been estimated that five hundred large planes flying two sorties a day for fifty days could spread a thin film of soot over the entire arctic ice cap. The absorption of solar energy by the black surface would – so the theory goes – melt the ice and cause climatic change favourable to northern lands such as Russia and Canada. The cost of this plan was computed to be about 2 billion dollars, a small price to pay compared with the investment in more conventional aggressive weapons of war.[37]

But there are other costs that might appear in the final reckoning. Water levels would rise, flooding coastal areas around the world. Changes in temperature differences between the polar regions and the equator would occur, disturbing the weather patterns and agricultural expectations of the whole planet. Everyone would suffer, the aggressors as well as the intended victims.

In the wake of the bitter cold that descended on the United States in the winters of 1977 and 78, many voices have been raised demanding that 'we do something about the weather'. Enthusiastic accounts in the press describe the weather-modification projects that are currently in use in many countries: rainmaking, hail suppression, fog dispersal. But the articles rarely mention the uncertainties surrounding the results of these relatively small-scale experiments. Our understanding of the forces that move the major climate systems of the earth is even more primitive. Scientists who have spent their whole lives on the subject do not agree on the significance of the changes that seem to be taking place today, let alone on what action could be taken to correct them.

Given the impetus of these extraordinary winters, however, the need for more intensive research in climatology may be recognized. In the face of a common enemy, it may be possible for nations to cooperate and to bring the study of climate modification out from behind the curtain of military secrecy. Acting together, the major producing and exporting countries of the world could plan for greater fuel demands and declining agricultural yields.

Eventually weather scientists may understand the complex cosmic rhythms that modulate the flow of sunlight on earth. They will be able to measure and predict more accurately the effect of altered air chemistry, of smog and volcanic dust, weaving all these factors into a coherent pattern. Then it may be possible to foresee and influence beneficially long-term changes. Until such understanding is achieved, any scheme to tamper with the weather could trigger chain reactions that we would be powerless to arrest. Man-made climate modifications designed to make the purga blow with less fury across Siberia might bring the simoom into lands that had never felt its abrasive touch, and turn the gentle 'sigh in the sky of China' into a chilling blast off the Gobi Desert. From the drought-stricken plains of Kansas to the dusty camps of Timbuktu, desperate people would compete for the remaining stores of food. Who knows what atrocities might be perpetrated in the shadow of world-wide hunger?

With or without deliberate human intervention, the climate of the earth cannot be expected to remain unchanged. It is apparent from a look at climatic history that winters centuries long descend upon the planet and sweltering summers last for aeons of time. A look at our sister planets, Venus and Mars, tells us that these cosmic seasons may not be simple repetitive cycles but waves in a long steady tide of change. Tomorrow will not be like yesterday, nor next year like the one that is past. The very substance of earth's aura is slowly being transmuted as the planet spins out its days, and circles year after year around its star, and moves in a vast arc around the glowing heart of the Galaxy. The earth itself and its cargo of living matter are continuously evolving and modifying each other as they move in their intricate path through space and time.

CHAPTER SIXTEEN

THE NINETY-YEAR FORECAST

Oh world, I cannot hold thee close enough!
Thy winds, thy wide gray skies!
Thy mists, that roll and rise!

– Edna St Vincent Millay[1]

Just a score of years has passed since Sputnik I soared in a flaming arc high above the windswept grasses of the Russian steppes, far above the cascading ice crystals of the mare's tails and the clouds that glow all night, to become the first man-made star in the night sky. In these brief years we have learned more about the nature of our home in space than in all the previous four million years of human existence on earth. For the first time we can see it whole, and we look with wonder on this variegated and beautiful bubble of matter framed in its aureole of blue. As Lewis Thomas so eloquently expresses it:

> Viewed from the distance of the moon, the astonishing thing about the earth, catching the breath, is that it is alive. The photographs show the dry, pounded surface of the moon in the foreground, dead as an old bone. Aloft, floating free beneath the moist, gleaming membrane of bright blue sky, is the rising earth, the only exuberant thing in this part of the cosmos. If you could look long enough, you would see the swirling of the great drifts of white cloud, covering and uncovering the half-hidden masses of land. If you had been looking for a very long geological time, you could have seen the continents themselves in motion, drifting apart on their crustal plates, held afloat by the fire beneath. It has the organized, self-contained look of a live creature, full of information, marvelously skilled in handling the sun.[2]

The earth's atmosphere, Thomas suggests, is like a membrane, the thin, pliable tissue which surrounds each living cell and enables the cell to maintain its identity while at the same time responding to its environment. The membrane acts as a selective filter, passing through useful

36. Earth's aura photographed from space, the moon riding high above.

substances, excluding harmful ones; so the cell can capture energy and nutrients in just the needed amounts. 'When the earth came alive,' Thomas says, 'it began constructing its own membrane for the purpose of editing the sunlight.'

Earth's aura is like a cellular membrane but even more sensitive, more complex. It is arranged in many layers, whose boundaries can actually be seen as a spaceship rises through the atmosphere. It is interlaced with rapid jet streams that flow like arteries carrying energy from one part of the system to another. Its prevailing winds are in constant motion, maintaining a dynamic exchange of life-giving solar heat around the globe, creating the ever changing patterns – the choreography of light and shadow, the orchestration of wind sounds and the voices of the sea. With its smooth lee waves, its *piles d'assiettes*, its radiant coronae and rainbows, earth's aura is as responsive as a chameleon's skin, as finely articulated as a butterfly's wing.

Responsiveness and sensitivity are characteristic of all animate creatures. Picking up tiny impulses from the environment, they magnify them in many remarkable ways. Human senses, for instance, depend upon electrical stimuli that trigger the nerve endings. Plants respond to light, opening their blossoms and turning their leaves to catch the sunshine. A sea anemone shrinks at an alien touch, drawing in its pale floating fronds. With similar sensitivity, the thin air of the upper troposphere reacts to the sharp, metallic probe of our jet planes, leaving long streaks of vapour across the sky. A single pellet of dry ice dropped into a formless grey cloud makes millions of snow crystals leap into lacy patterns of intricate form. And high above the planet where the wind from the sun turns and spirals in the earth's magnetic field, signals (not yet deciphered by man) may cause winds to gather on earth and tornadoes to dangle their dark snuffing trunks above the wheat fields and the cities of the Great Plains.

Every living thing secretes special organic chemicals that regulate vital processes – insulin, for example, controls the level of sugar in the blood. In a similar way chlorophyll draws in the energy of sunlight, stores it for later use, and keeps the earth's air supplied with oxygen.

Many feedback processes also initiate and sustain periods of progressive change in living, growing things. In the earth's atmosphere a temperature drop of a degree or two may precipitate an accelerating advance of the polar ice caps. A tiny cloud of hot air rising from a tropic sea can start to spin and grow until it becomes a raging whirlwind of destruction.

So the earth responds to the many diverse influences that flow in upon it from outer space or emanate from the curious assortment of living creatures that inhabit its crust. It selects, rejects, alters, and magnifies them in creative ways as though it had a vital force and a will of its own.

The aura itself – that ring of tenuous matter suffused with energy – is in some mystical way an essential characteristic of being alive. 'Life is a luminous halo,' says Virginia Woolf, 'a semi-transparent envelope surrounding us from the beginning of consciousness to the end.'[3]

Although the earth has many attributes of life, most biologists, I suspect, would be loath to admit it to the special company of living things. But I am not too concerned about their scepticism. When they can clearly define the difference between life and non-life, then the question can be debated. In the meantime, it would be useful to think of the planet earth, surrounded by its halo of gases, as though it were a living organism – to sense its changing needs, to protect its flow of sunshine, bind its wounds, and treat it with the same loving gentleness we accord to a flower or a child.

'One thing alone life does not appear to do,' Loren Eiseley says; 'it never brings back the past. Unlike lifeless matter, it is historical. It seems to have had a single origin and to be traveling in a totally unique fashion in the time dimension.'[4]

What is this future towards which we are travelling, our planet with its envelope of individual organisms held in symbiotic embrace? With the aid of computers we can now predict with reasonable confidence the condition of the earth's atmosphere one day ahead, even ninety days in the future. But ninety days is not long enough to allow us to plan how best to meet the changing conditions of our world. We need a span of time more closely matching a human lifespan. Fourscore or ninety years, perhaps. The explosive growth of knowledge in the last two decades lends substance to the belief that in ninety years many of the major questions about the forces that influence the earth's atmosphere could be resolved and ways could be found to promote conditions that would be favourable not only to ourselves but also to the other creatures that share this cosmic adventure with us.

Ninety years is less than one grain of sand in the hourglass of geological time; yet – given the power of technology – it is long enough to alter the face of the planet. It is long enough to turn the 'temperate' regions of the earth into a tropical jungle or to cover the high latitudes with a sheet of ice. It is long enough to fray the thin fabric of the ozone umbrella so it can no longer hold off the steady rain of ultra-violet light.

Ninety years is just one generous human lifetime and many babies born today would live to see a ninety-year forecast come true. But even the most sophisticated computers available today cannot make a prediction that far into the future. There are too many unknown factors. The most important one is man himself. Only a little while ago he was powerless to make the earth tremble or to alter even for an instant the routine of her appointed rounds. All that is changed; man now makes clouds form and rains fall; he deflects the path of hurricanes and the flow of volcanic lava; he dreams of melting the polar caps with a layer of soot; he builds cities that make hail form and he puts a veil of dust between himself and the sun.

From this time on, the texture of earth's aura, the destiny of the planet itself, is linked inextricably with the nature of man. *Homo sapiens* – the thinking creature as distinguished from all other species. Is this really the central core of man's nature? How strong is his greed? How deep his drive for power? In fourscore years the answer will be blowing in the wind. Today it still lies hidden in the heart and the mind of man.

NOTES

Introduction. The View from Space

1. *Epigraph*: Astronaut in Apollo 8, as quoted by Anne Morrow Lindbergh in *Earth Shine*, Harcourt Brace, New York, 1969.

2. *Gagarin quotation*: as quoted by K. Ya. Kondratyev and O. I. Smokty, 'Twilight Seen from Space', *Natural History*, Vol. 82, No. 9, November 1973, p. 12.

Chapter 1. Man Probes the Atmosphere

1. *Epigraph*: Plato, *Phaedra*, Jowett translation, as quoted by David I. Blumenstock in *The Ocean of Air*, Rutgers University Press, New Brunswick, N.J., 1959.

2. *John Wise and Nellie*: for the facts in this portion I followed the description of Kurt R. Stehling and William Beller in *Skyhooks*, Doubleday & Company, Garden City, N.Y., 1962, pp. 54–78.

3. *Wise quotation*: as reported by Stehling and Beller, *Skyhooks*, p. 62.

4. *Galileo and Torricelli's experiment*: as described by James B. Conant, *Science and Commonsense*, Yale University Press, New Haven, Conn., 1951, pp. 69–73.

5. *Pascal quotations*: from *The Physical Treatise of Pascal*, trans. by I. H. B. and A. G. H. Spiers, Columbia University Press, New York, 1937.

6. *Périer's letter*: from *The Physical Treatise of Pascal*.

7. *Zambeccari quotation*: from excerpt in Glines, *Lighter-than-Air Flight*, pp. 46–7.

8. *Montgolfier, Zambeccari, etc.*: for the facts of the early history of ballooning I have relied chiefly upon Abraham Wolf, *History of Science, Technology, and Philosophy*, Macmillan Company, New York, 1950; and C. V. Glines, ed., *Lighter-than-Air Flight*, Franklin Watts, New York, 1965.

9. *Glaisher quotation*: from J. Glaisher, C. Flammarion, W. Fonvielle, G. de Tissandier, 'Coxwell and Glaisher's Dangerous Ascent', *Voyages Aériens*, Paris, 1870.

10. *Glaisher and Coxwell flight*: for the details of this flight I have followed very closely the story as told by Stehling and Beller in *Skyhooks*. This book contains many fascinating descriptions of balloon flight and is highly recommended as additional reading on this subject.

11. *First Simons quotation*: David G. Simons with Don A. Schanche, *Man High*, Doubleday & Company, Garden City, N.Y., 1960, p. 135.

12. *Second Simons quotation:* David G. Simons, 'Man High', *Life*, 21 October 1957, pp. 19–27.

13. *Older theory of distribution of gases in the atmosphere:* proposed by Edmund Halley in 1714, as reported by David I. Blumenstock in *The Ocean of Air*, Rutgers University Press, New Brunswick, N.J., 1959, pp. 163–4.

Chapter 2. Water Falling from the Sky

1. *Epigraph:* Elinor Wylie, 'Velvet Shoes', in *Collected Poems*, Alfred A. Knopf, New York, 1932.

2. *New York Times description:* as quoted by Lowell Thomas in *Hungry Waters*, John C. Winston Co., Chicago, 1938, p. 118.

3. *Amount of water in atmosphere, seas, etc.:* for these facts I have used the information given by H. L. Penman, 'The Water Cycle', *Scientific American*, Vol. 223, No. 3, September 1970, p. 102.

4. *Dew pits:* described by Paul Sears in *Where There Is Life*, Dell Publishing Co., New York, 1962, p. 145.

5. *Fog near the seashore:* figures quoted by Blumenstock, *The Ocean of Air*, p. 114.

6. *Eiseley quotation:* from Loren Eiseley, 'The Flow of the River', *The Immense Journey*, Vintage Books, Alfred A. Knopf and Random House, New York, 1957, p. 27.

7. *Snow in Finland:* cited by Guy Murchie, *Song of the Sky*, Riverside Press, Cambridge, Mass.; Houghton Mifflin Co., Boston, 1954, p. 249.

8. *Green light in hailstorms:* as reported to me by T. Theodore Fujita, University of Chicago, in a personal interview.

9. *Hailstorms in India and France:* cited by Leo Loebsack in *Our Atmosphere*, trans. by E. L. and D. Rewald, New American Library, New York, 1961, p. 124.

10. *Anti-hail bell ringing and guns:* for these historical facts I have relied upon D. S. Halacy, Jr, *The Weather Changers*, Harper & Row, New York, 1968, p. 129.

11. *Soviet successes with hail reduction:* McGraw-Hill Encyclopedia of Science and Technology, Vol. 14, McGraw-Hill Book Co., New York, 1971, p. 524.

12. *Schaefer, Langmuir, and Vonnegut experiments:* for the details of this history I have relied upon J. Gordon Cook's account in *Our Astonishing Atmosphere*, Dial Press, New York, 1957, pp. 128–9.

13. *Snow produced by a single dry ice pellet:* as cited by Cook in *Our Astonishing Atmosphere*, p. 131.

14. *Hail study at NCAR:* as reported in 'Hail Suppression Up in the Air', *Science*, Vol. 191, 5 March 1976, p. 932.

15. *Weather-modification projects:* as reported by Nigel Calder in *The Weather Machine*, Viking Press, New York, 1974, pp. 83–4.

16. *Measurements of high cloud cover:* as reported in 'Is Man Changing the Earth's Climate?', *Chemical and Engineering News*, 16 August 1971, pp. 40–41.

17. *La Porte weather changes:* as cited in the *McGraw-Hill Encyclopedia of Science and Technology*, Vol. 3, p. 185.

18. *St Louis experiment:* as reported by Roscoe R. Braham, Jr, in 'Overview of Urban Climate', *Proceedings, Conference on Urban Physical Environment*, Syracuse, N.Y., 25–9 August 1975, and in 'Cloud Physics of Urban Weather Modification – A Preliminary Report', *Bulletin of the American Meteorological Society*, Vol. 55, No. 2, February 1974, pp. 100–106.

Chapter 3. Spindrift of the Ocean of Air

1. *Epigraph:* Henry David Thoreau, *Journal*, Houghton Mifflin, Boston, 1949.

2. *Clustering of cumulus clouds around the equator:* my account of the satellite pictures and frequency of clouds near the Marshall Islands is based on Calder's description in *The Weather Machine*, pp. 50–52.

3. *Pückler-Muskau quotation:* Prince Pückler-Muskau, 'Frail Aerial Bark', excerpted in Glines, ed., *Lighter-than-Air Flight*, pp. 49–50.

4. *Loebsack quotation:* from Loebsack, *Our Atmosphere*, transl by E. L. and D. Rewald, New American Library, New York, 1961, p. 122.

5. *Simons quotation:* from Simons, *Man High*, p. 168.

6. *Balls of lightning:* as described and 'explained' in James R. Newman, ed., *Harper Encyclopedia of Science*, Harper & Row, New York; Sigma, Washington D.C., 1967, p. 134.

7. *Estimate of ships struck by lightning:* as reported by Loebsack, *Our Atmosphere*, p. 129.

8. *Aborigines of Tasmania:* custom of gathering burning faggots from lightning described by Blumenstock in *The Ocean of Air*, p. 391.

9. *Franklin's letter:* from Forest Ray Moulton and Justus J. Schifferes, eds., *The Autobiography of Science*, 2nd ed.; Doubleday & Company, Garden City, N.Y., 1960, p. 236.

10. *Incidence of thunderstorms around the world:* figures cited by T. J. Chandler, *The Air Around Us*, Natural History Press for the American Museum of Natural History, Garden City, N.Y., 1969, p. 95.

11. *Simons quotation:* from Simons, *Man High*, p. 155.

12. *Undset quotation:* as quoted by Loebsack, *Our Atmosphere*, p. 36.

13. *Mother-of-pearl clouds and Halley's comet:* this sighting was described by

M. Minnaert in *Light and Colour in the Open Air*, trans. by H. M. Kremer-Priest; rev. by K. E. Brian Jay, G. Bell & Sons, London, 1940, p. 230.

14. *El Fuego dust cloud:* the information on the N A S A experiments came from personal communication with William H. Fuller, Jr, at N A S A Langley Research Center, Hampton, Va. The earlier measurements are also cited in M. P. McCormick and W. H. Fuller, Jr, 'Lidar Measurements of Two Intense Stratospheric Dust Layers', *Applied Optics*, Vol. 14, No. 1, January 1975, p. 4.

15. *Noctilucent clouds:* for my description of these clouds I relied upon Minnaert, *Light and Colour in the Open Air*, pp. 284–6.

16. *Noctilucent clouds and the Siberian meteor:* this sighting was reported by Minnaert, *Light and Colour in the Open Air*, p. 286.

17. *Experiment in Sweden on noctilucent clouds:* for the details of this experiment I have relied upon Robert K. Soberman's article, 'Noctilucent Clouds', *Scientific American*, Vol. 208, No. 6, June 1963, pp. 51–9.

Chapter 4. Cycles in the Biosphere

1. *Epigraph:* John Hall Wheelock, 'This Quiet Dust', *Dust and Light*, Charles Scribner's, New York, 1919.

2. *Priestley's experiment:* for the details of this scene I have relied upon the description given by Bernard Jaffe in *Crucibles*, Tudor Publishing Co., New York, 1930, pp. 51–72.

3. *Priestley quotation:* from Joseph Priestley, *Experiments and Observations on Different Kinds of Air* (1775), as excerpted in Dagobert D. Runes, ed., *Treasury of World Science*, Philosophical Library, New York, 1962, p. 864.

4. *Priestley quotation:* from Joseph Priestley, *Experiments and Observations on Different Kinds of Air* (1775), as excerpted in Moulton and Schifferes, eds., *The Autobiography of Science*, 2nd ed., Doubleday & Co., Garden City, N.Y., 1960, p. 226.

5. *Estimate of depth required for protection:* cited by Lewis Thomas in *The Lives of a Cell*, Viking Press, New York, 1974, p. 146.

6. *Sensitivity of organisms to oxygen:* for these estimates I have relied upon the information given by Preston Cloud and Aharon Gibor in 'The Oxygen Cycle', *Scientific American*, Vol. 223, No. 3, September 1970, pp. 110–23. Louis Pasteur's estimate was cited in this article.

7. *Eiseley quotation:* from Loren Eiseley, 'How Flowers Changed the World', *The Immense Journey*, Vintage Books, Alfred A. Knopf and Random House, New York, 1957, p. 76. Eiseley's description of the explosive evolution of angiosperms was the basis of my remarks on this subject.

8. *Biological effects of carbon dioxide:* I have used the effects cited by Loebsack in *Our Atmosphere*, p. 135.

9. *Fluctuations in carbon dioxide:* for the information in this section I have relied on the figures given by Bert Bolin in 'The Carbon Cycle', *Scientific American*, Vol. 223, No. 3, September 1970, pp. 125–32.

10. *Amount of nitrogenase in the world:* estimate as cited by C. C. Delwiche in 'The Nitrogen Cycle', *Scientific American*, Vol. 223, No. 3, September 1970, p. 142 ('probably no more than a few kilograms').

Chapter 5. Man's Aura

1. *Epigraph:* Shakespeare, *Hamlet*, Act II, Scene 2.

2. *Quotation from the Evening News:* as quoted by William Wise in *Killer Smog*, Rand McNally & Company, Chicago, 1968, p. 60.

3. *Evelyn quotation:* from John Evelyn, *Fumifugium, or, The Inconvenience of the Air, and Smoke of London Dissipated. Together with Some Remedies humbly proposed by J. E. Esq. to his Sacred Majesty, and to the Parliament now Assembled,* as excerpted in Wise, *Killer Smog*, p. 21.

4. *London smog of 1952:* for the details of this pollution episode I have relied upon the description given by Louis J. Battan in *The Unclean Sky*, Doubleday & Company, Garden City, N.Y., 1966, pp. 1–4, and by Wise in *Killer Smog*. The stories of the lost swan and the Southwark morgue were experiences related to Wise by survivors of the London smog.

5. *Medical journal quoted:* as quoted by Wise in *Killer Smog*, p. 145 (journal not identified).

6. *Ozone in smog:* A. J. Haagen-Smit, 'The Control of Air Pollution', *Scientific American*, Vol. 210, No. 1, January 1964, pp. 25–31.

7. *Ozone injury to vegetation:* U.S. Department of Health, Education and Welfare, *Air Quality Criteria for Photochemical Oxidants*, National Air Pollution Control Administration Publication No. AP-63, Washington, D.C., 1970, pp. 6:8, 6:9; also George D. Clayton *et al.*, 'Community Air Quality Guides, Ozone (photochemical oxidant)', *American Industrial Hygiene Association Journal*, May–June 1968, p. 301; and Leon S. Dochinger, 'Christmas Trees and Air Pollution', *Horticulture*, December 1973.

8. *Studies at Yale University on rate of photosynthesis:* 'Effect of Ozone on Photosynthesis', *Bulletin of the Ecological Society of America*, Vol. 51, No. 2, June 1970.

9. *Results of ozone exposure on laboratory animals and human volunteers:* Clayton *et al.*, 'Community Air Quality Guides', p. 300. Also U.S. Department of Health, Education and Welfare, *Air Quality Criteria for Photochemical Oxidants*, pp. 8:33 and 10:5 to 10:8.

10. *Ozone in Los Angeles:* data cited by Mariana Gosnell, in 'Ozone – The Trick Is Containing It Where We Need It', *Smithsonian*, June 1975, p. 50.

11. *Ozone in Chicago:* data provided by the City of Chicago Department of Environmental Control.

12. *Ozone from power plants:* D. D. Davis, G. Smith, and G. Klauber, 'Trace Gas Analysis of Power Plant Plumes via Aircraft Measurement: O_3, NO_x, and SO_2 Chemistry', *Science*, Vol. 186, 22 November 1974, pp. 733–5.

13. *Acid rain:* my information about acid rain has been drawn from the following articles: 'Scientists Puzzle over Acid Rain', *Chemical and Engineering News*, 9 June 1975, pp. 19–20; and 'Norway: Victim of Other Nations' Pollution', *Chemical and Engineering News*, 14 June 1976, pp. 15–16.

14. *Sulfur dioxide and acid rain in New York State:* measurements cited by Gene E. Likens and F. Bormann, 'Acidity in Rainwater: Has an Explanation Been Presented?', *Science*, Vol. 188, 30 May 1975, pp. 957–8.

15. *Remark of official:* as quoted in *Newsweek*, 20 March 1972, p. 8.

16. *Fish in Norway:* research reported in 'Norway: Victim of Other Nations' Pollution'.

17. *Scientists photographing auras:* reported in Stanley Krippner and David Ruben, eds., *The Kirlian Aura: Photographing the Galaxies of Life*, Anchor Press/Doubleday, Garden City, N.Y., 1974.

Chapter 6. The Ozone Umbrella

1. *Epigraph:* Matthew Arnold, 'Stanzas in Memory of the Author of "Obermann"'.

2 *Airglow:* for the facts on this phenomenon I have relied upon M. F. Ingham's article, 'The Spectrum of the Airglow', *Scientific American*, Vol. 226, No. 1, January 1972, pp. 78–85.

3. *Zelikoff experiment:* as described by Loebsack in *Our Atmosphere*, pp. 157–60.

4. *Potential of this man-made moonlight:* possibilities for exploitation as expressed by Loebsack in *Our Atmosphere*, p. 160.

5. *Variations in density of ozone layer:* as cited in 'Fluorocarbons and the Environment', report of Federal Task Force on Inadvertent Modification of the Stratosphere, Council on Environmental Quality and Federal Council for Science and Technology, June 1975, p. 69.

6. *Statistics from survey of National Cancer Institute:* as cited by Ogden Tanner in 'The Great Spray-Can Scare', *Nature/Science Annual 1976*, Time-Life Books, New York, 1976, p. 118. See also 'The Possible Impact of Fluorocarbons and Halocarbons on Ozone, May 1975', Interdepartmental Committee for Atmos-

pheric Sciences, Federal Council for Science and Technology Policy Office, National Science Foundation, May 1975, pp. 65–75.

7. *Projection of increased cases of skin cancer:* estimates taken from 'The Possible Impact of Fluorocarbons and Halocarbons on Ozone, May 1975', p. 56.

8. *Incidence of malignant melanoma:* figures and opinions as cited by Michael B. McElroy in 'Threats to the Atmosphere', *Harvard Magazine*, February 1976, pp. 19–25. Also 'Fluorocarbons and the Environment', p. 72.

9. *Effects on plankton and other organisms:* studies cited in 'Fluorocarbons and the Environment', pp. 53–64.

10. *Thickness of ozone layer if brought down to sea level:* Figure of 3 millimetres cited by J. A. Ratcliffe in *Sun, Earth, and Radio*, McGraw-Hill Book Co., New York, 1970, p. 25.

11. *Quotation from report: Environmental Impact of Stratospheric Flight*, Report of the Climatic Impact Committee of the National Research Council, the National Academy of Sciences, and the National Academy of Engineering, National Academy of Sciences, Washington, D.C., 1975, p. 6.

12. *Farrell quotation:* from 'The War Department Release on the New Mexico Test', 16 July 1945.

13. *Effect of nuclear attack:* as reported by the National Academy of Sciences, 'Long-Term Worldwide Effects of Multiple Nuclear-Weapon Detonations', 1975, pp. 6–7.

14. *Effect of jet engines on ozone layer:* for the information in this section I relied upon the report published by the National Academy of Sciences, *Environmental Impact of Stratospheric Flight*, pp. 13–15.

15. *M.I.T. Study:* Fred N. Alyea, Derek M. Cunnold, and Ronald G. Prinn, 'Stratospheric Ozone Destruction by Aircraft-Induced Nitrogen Oxides', *Science*, Vol. 188, 11 April 1975, p. 120.

16. *Effect of each Boeing 2707:* probable increase in skin cancer incidence cited in *Environmental Impact of Stratospheric Flight*, p. 18.

17. *Quote from report of National Academy of Sciences: Environmental Impact of Stratospheric Flight*, pp. 14–15.

18. *Amount of fluorocarbons expelled annually:* figure cited in 'Fluorocarbons and the Environment', p. 2.

19. *Diffusion and residence time for chlorine atoms:* figures cited in 'Chloro-fluorocarbons Threaten Ozone Layer', *Chemical and Engineering News*, 23 September 1974, pp. 28–9.

20. *Fluorocarbon found above 12 miles:* reported in 'Chlorofluorocarbon-Ozone Issue Flares Up Again', *Chemical and Engineering News*, 11 August 1975, p. 13.

21. *Hydrogen chloride measurements:* reported by Walter Sullivan in 'Climatic

Changes by Aerosols in Atmosphere Feared', *The New York Times*, 14 September 1975, p. 1.

22. *Estimates in time delay in ozone layer depletion:* see Steven C. Wofsy, Michael B. McElroy, and Nien Dak Sze, 'Freon Consumption: Implications for Atmospheric Ozone', *Science*, Vol. 187, 14 February 1975, p. 536. Also *Halocarbons: Effects on Stratospheric Ozone*, report of Panel on Atmospheric Chemistry, Committee on Impacts of Stratospheric Change, National Academy of Sciences, Washington, D.C., September 1976, p. 1–23.

23. *Formation of chlorine nitrate:* reported in *Chemical and Engineering News*, 31 May 1976, p. 14, and 5 July 1976, p. 7. Also *Halocarbons: Effects on Stratospheric Ozone*, pp. 1–13 and 1–14.

24. *Less destructive type of fluorocarbon:* as described in 'Scientists Detail Chlorofluorocarbon Research', *Chemical and Engineering News*, 21 April 1975, p. 23. FC-22 is said to be twenty times less hazardous than FC-12. See also 'Fluorocarbons and the Environment', p. 95.

25. *Fluorocarbons in refrigeration:* estimates cited by Michael B. McElroy in 'Threats to the Atmosphere', p. 22.

26. *Quote from Du Pont statement: Quarterly Report and Semiannual Statement*, Second Quarter, 1975, E. I. Du Pont de Nemours and Company, p. 11.

27. *Latency period for skin cancer:* as cited in 'Fluorocarbons and the Environment', p. 72.

28. *National Academy of Sciences report: Halocarbons: Effects on Stratospheric Ozone*, pp. 3–9 and 3–11. Also *Halocarbons: Environmental Effects of Chlorofluoromethane Release*, report of Committee on Impacts of StratoChange, National Academy of Sciences, Washington, D.C., September 1976, pp. 1–9.

29. *Phased end to fluorocarbon uses:* Statement by U.S. Food and Drug Administration, Environmental Protection Agency, and Consumer Product Safety Commission in *Federal Register*, 13 May 1977. Regulations set by the three federal agencies, *Federal Register*, 17 March 1978.

Chapter 7. The Wind's Way

1. *Epigraph:* Dante Gabriel Rossetti, 'Woodspurge'.

2. *Villiers quotation:* from Alan Villiers, *Wild Ocean*, McGraw-Hill Book Co., New York, 1957, pp. 12–16.

3. *Coleridge quotation:* from Samuel Taylor Coleridge, *The Rime of the Ancient Mariner*.

4. *Columbus quotation: Columbus' Journal of His First Voyage to America* as quoted by A. B. C. Whipple in *Tall Ships and Great Captains*, Harper & Brothers, New York, 1951, pp. 28–9.

5. *Distribution of land birds:* as reported by Carl Welty in 'The Geography of Birds', *Scientific American*, Vol. 197, No. 1, July 1957, p. 118.

6. *Murchie quotation:* from Murchie, *Song of the Sky*, pp. 176–7.

7. *Calculation of human 'spider web':* as reported in 'Phenomena, Comment, and Notes', *Smithsonian*, October 1975.

8. *Forms of life on Krakatoa:* these observations were described by Rachel Carson, *The Sea Around Us*, Oxford University Press, New York, 1951, pp. 90–92.

9. *Climate in North Africa in Roman times:* for this information I have relied upon Slater Brown's account in *World of the Wind*, Bobbs-Merrill Company, Indianapolis, 1961, pp. 45–6.

10. *Dana quotation:* from Richard Henry Dana, Jr, *Two Years Before the Mast*, Heritage Press, New York, 1947; orig. pub. 1866, excerpts from Chs. 31–2.

Chapter 8. The Swift Jet Streams

1. *Epigraph:* Shakespeare, *Hamlet*, Act I, Scene 5.

2. *Japanese balloon found in Montana:* for the details of this episode I have drawn on the information published by *Time*, 1 January 1945, p. 14; and *Newsweek*, 1 January 1945, p. 36, and 15 January 1945, p. 40.

3. *Japanese balloon offensive:* for the facts on this subject I have relied upon two sources: Lincoln LaPaz, with Albert Rosenfeld, 'Japan's Balloon Invasion of America', *Collier's*, 17 January 1953, pp. 9–11; and Brig. Gen. W. H. Wilbur (USA), 'Those Japanese Balloons', in Glines, ed., *Lighter-than-Air Flight*.

4. *Discovery of jet stream by American pilots:* as described by Elmer R. Reiter in 'Rapid Rivers of Air', *Natural History*, Vol. 84, No. 3, March 1975, pp. 46–51.

5. *Mason quotation:* from Monck Mason, *Aeronautica; or Sketches Illustrative of the Theory and Practice of Aerostation: Comprising an Enlarged Account of the Late Aerial Expedition to Germany*, F. C. Westley, London, 1838, pp. 130–31. I was led to the Mason quotation by Murchie, who used it in *Song of the Sky*.

6. *Jet streams may help to spawn tornadoes:* this theory was mentioned in 'News from the World of Space Exploration', *Space World*, October 1970, pp. 43–6. It is not, however, accepted by all tornado experts.

7. *Weather satellites and balloons:* the information in this section is based upon Calder's account in *The Weather Machine*, pp. 57–62.

Chapter 9. The Whirlwinds

1. *Epigraph:* Lafcadio Hearn, as quoted by Fairfax Downey in *Disaster Fighters*, G. P. Putnam's Sons, New York, 1938.

2. *Sea Islands hurricane:* my description is based upon the account in Fairfax Downey's *Disaster Fighters*, G. P. Putnam's Sons, New York, 1938, pp. 47–55; and Joel Chandler Harris, 'The Sea Islands' Hurricane', *Scribner's Magazine*, February and March 1894, pp. 247–9. *Scribner's* sent Harris to the Sea Islands to interview the survivors.

3. *Quote from a survivor of the storm:* as reported by Downey in *Disaster Fighters*, G. P. Putnam's Sons, New York, 1938, p. 55. Spelling of dialect has been modernized.

4. *Conrad quotation:* from 'Typhoon' by Joseph Conrad in *Typhoon and Other Stories of the Sea*, Dodd, Mead & Company, New York, 1963, pp. 71–2.

5. *'Low pressures strained their eardrums':* as reported by Loebsack in *Our Atmosphere*, p. 112.

6. *Floods in the Bay of Bengal:* as described by Loebsack in *Our Atmosphere*, p. 112; and R. Cecil Gentry in 'Disaster from the Tropics', *Natural History*, Vol. 82, No. 8, October 1973, p. 47.

7. *Hurricanes strike with a force equal to 1,000 atomic bombs:* based on Andrew H. Brown's statement in 'Man against the Hurricane', *National Geographic*, October 1950, p. 537.

8. *Hurricane reconnaissance planes:* my description of the hurricane watch system in the 1940s and 1950s is based on Andrew H. Brown's 'Man against the Hurricane', and C. F. Tannehill's *The Hurricane Hunters*, Dodd, Mead & Company, New York, 1956.

9. *Tropical storms off west coast of Mexico:* reported in *Space World*, January 1975, pp. 32–3.

10. *Damage from hurricane Celia:* reported in *Science News*, 21 August 1971, p. 129.

11. *Hazards of hurricane modification:* see, for example, R. A. Howard, J. E. Matheson, and D. W. North, 'The Decision to Seed Hurricanes', *Science*, Vol. 176, 16 June 1972, pp. 1191–5.

12. *Keller quotation:* as quoted by Brown in *World of the Wind*, pp. 199–200.

13. *Levy's encounter with tornado:* as reported in *Newsweek*, 23 September 1957, pp. 102–3.

14. *Statistics of tornadoes in 1972:* based on figures reported by Louis J. Battan in 'Killers from the Clouds', *Natural History*, Vol. 84, No. 4, April 1974, p. 57.

15. *Electrical theory of tornado origin:* described by R. T. Ryan and B. Vonnegut in 'Eyewitness Tornado Observations Obtained with Telephone and Tape Recorder', *Weatherwise*, Vol. 23, June 1970, p. 126–30.

16. *Tornadoes bred by unusually high thunderstorms:* theory suggested by T. Theodore Fujita in 'Proposed Mechanism of Tornado Formation from Rotating Thunderstorm', *Proceedings of the Eighth Conference on Severe Local Storms, October 15–17, 1973,* American Meteorological Society, Boston.

17. *Winds cutting across the tops of thunderheads:* see reference note 6 for Chapter 8, *Jet streams may help to spawn tornadoes.*

18. *Hook-shaped echoes on radar:* described by Fujita, 'Proposed Mechanism of Tornado Formation from Rotating Thunderstorm', *Proceedings of the Eighth Conference on Severe Local Storms, October 15–17, 1973,* American Meteorological Society, Boston.

19. *Incidence and location of tornadoes:* these facts were cited by T. Theodore Fujita, Allen Pearson, and David Ludlum in 'Long-Term Fluctuation of Tornado Activities', *Proceedings of the Ninth Conference on Severe Local Storms, October 21–23, 1975, Norman, Oklahoma,* American Meteorological Society, Boston, pp. 417–23. Most recent figures by personal communication.

20. *45-year cycle:* theory proposed by Fujita, Pearson, and Ludlum in 'Long-Term Fluctuation of Tornado Activities'.

21. *Tornado incidence around the world:* these facts are based on information given by T. Theodore Fujita in 'Tornadoes Around the World', *Weatherwise,* Vol. 26, No. 2, April 1973, pp. 56–83.

22. *Tornadoes in and around cities:* based on information given by Fujita in 'Tornadoes Around the World'. Tornadoes caused by the earthquake and fires in Tokyo were described in this article.

Chapter 10. Lee Waves, Williwaws, and Ripples of Sand

1. *Epigraph:* Jerome Lawrence and Robert E. Lee, *The Night Thoreau Spent in Jail.*

2. *Shower of blood in Paris:* as reported by Brown in *World of the Wind,* Bobbs-Merrill Company, Indianapolis, 1961, p. 38.

3. *Reduction of wind force by passage through a forest:* experiments mentioned by Brown in *World of the Wind,* p. 46, citing the *Yearbook of Agriculture.*

4. *127 names for regional winds:* collected by Murchie as described in *Song of the Sky,* pp. 131–6.

5. *Winds on Mount Washington:* wind speeds cited by Blumenstock in *The Ocean of Air,* p. 88.

6. *Shklovsky quotation:* from I. W. Shklovsky, *In Far North-East Siberia,* trans. by L. Edwards and Z. Shklovsky, Macmillan & Co., London, 1916, pp. 114–15. I was led to this quotation by Blumenstock, who used it in *The Ocean of Air.*

7. *Winds off the desert:* my description is based upon that given by Brown in *World of the Wind*, p. 37.

8. *Caravans in sandstorms:* as described by Brown in *World of the Wind*, p. 37.

9. *Nearly 80 per cent of land lies near a large body of water:* this statement is after Blumenstock, *The Ocean of Air*, p. 55.

10. *Brown quotation:* from Brown, *World of the Wind*, p. 84.

11. *Slocum quotation:* from Joshua Slocum, *Sailing Alone Around the World*, Appleton Century, New York, 1935; reprint 1954, pp. 105–6.

12. *Föhn winds:* the information in this section is based upon Loebsack's *Our Atmosphere*, pp. 98–9, and Brown's *World of the Wind*, pp. 66–8. Theory of the origin of föhn winds as described by Brown.

13. *'Leaves crumble to dust . . .':* quotation from R. DeC. Ward, *The Climates of the United States*, as quoted by Brown in *World of the Wind*, p. 82.

14. *The chinook and the temperatures in western Canada:* as cited by Brown in *World of the Wind*, pp. 78–9.

15. *Lee waves:* in describing this phenomenon I have relied upon R. S. Scorer, 'Lee Waves in the Atmosphere', *Scientific American*, Vol. 204, No. 3, March 1961, pp. 123–34.

16. *Philip Wills' glider flights near Mount Cook:* as described in Scorer, 'Lee Waves in the Atmosphere'.

17. *Wills quotation:* from Philip A. Wills, *The Beauty of Gliding*, Pitman Publishing Co., New York, 1960, as excerpted in Joseph Colville Lincoln, ed., *Soaring on the Wind*, Northland Press, Flagstaff, Arizona, 1972, p. 60.

18. *Une pile d'assiettes:* as described by Scorer in 'Lee Waves in the Atmosphere'.

19. *Larry Edgar's flight into the rotor cloud:* as recounted by Larry Edgar in 'Frightening Experience during Jet-Stream Project', *Soaring*, July–August 1955, pp. 20–23.

20. *Edgar quotation:* from Edgar, 'Frightening Experience during Jet-Stream Project'.

21. *Damage to Edgar's glider:* as assessed by Dr Joachim P. Kuettner and Lloyd Licher; see Joseph Colville Lincoln, ed., *On Quiet Wings*, Northland Press, Flagstaff, Arizona, 1972, p. 372.

Chapter 11. The Green Flash and Other Light Shows

1. *Epigraph:* Omar Khayyám, 'The Rubaiyat', translated by Edward Fitzgerald.

2. *Newton's comment about himself:* as quoted in Forest Ray Moulton and

Justus J. Schifferes, eds., *The Autobiography of Science*, 2nd ed., Doubleday & Co., Garden City, N.Y., 1960, p. 172.

3. *Newton's description of his experiment:* from Isaac Newton, *Principia Mathematica*, ed. by Florian Cajori from trans. by Andrew Motte, University of California Press, Berkeley, 1934.

4. *Color perception of insects and birds:* as reported by Georgii A. Mazokhin-Porshnyakov, *Insect Vision*, trans. by Roberto and Lilliana Masironi, Plenum Press, New York, 1969. Also R. H. Smythe, *Animal Vision: What Animals See*, Herbert Jenkins, London, 1961.

5. *The meaning of the rainbow in ancient cultures:* for this information I have relied upon Carl B. Boyer's account in 'Art, Myth, and Magic', in *The Rainbow Book*, Fine Arts Museums of San Francisco, San Francisco, 1975.

6. *Biblical quotation:* Authorized King James Version, Genesis 9, 12–15.

7. *Descartes's explanation of the rainbow:* from *Les Météors* (1637) by René Descartes as excerpted in Dagobert Runes, ed., *Treasury of World Science*, Philosophical Library, New York, 1962, p. 239.

8. *Full circle of rainbow seen from plane:* as described by Rutherford Platt in 'What Is a Rainbow?', *The Rainbow Book*, p. 77.

9. *Sunsets enhanced by volcanic cloud:* this effect was described by Minnaert in *Light and Colour in the Open Air*, pp. 277–80.

10. *Sunsets and the El Fuego dust cloud:* as reported by Aden B. and Marjorie P. Meinel in 'Stratospheric Dust-Aerosol Event of November 1974', *Science*, Vol. 188, 2 May 1975, pp. 477–8; also *Nature/Science Annual*, 1976, p. 177.

11. *Russian astronauts' observations:* from Kondratyev and Smokty, 'Twilight Seen from Space', p. 12.

12. *Green flash:* phenomenon as described by D. J. K. O'Connell, S.M., in 'The Green Flash', *Scientific American*, Vol. 202, No. 1, January 1960, pp. 112–22.

13. *Simons quotation:* from Simons, *Man High*, p. 172.

14. *Duration of green flash in Scandinavia and Antarctica:* as reported by O'Connell in 'The Green Flash', p. 120.

15. *Evershed's observations:* as described by O'Connell in 'The Green Flash', p. 120.

16. *Discovery of Iceland and Greenland:* the theory that this discovery might have been aided by the arctic mirage is presented by H. L. Sawatzky and W. H. Lehn in 'The Arctic Mirage and the Early North Atlantic', *Science*, Vol. 192, 25 June 1976, pp. 1300–1305.

17. *Peary's sighting of Crocker Land:* based upon the account given by Alistair B. Fraser and William H. Mach in 'Mirages', *Scientific American*, Vol. 234, No. 1, January 1976, p. 102.

18. *Expedition to Crocker Land:* this story is based upon the report of Donald B. MacMillan, 'Geographical Report of the Crocker Land Expedition, 1913–1917', *Bulletin of the American Museum of Natural History*, Vol. 56, Art. VI, 1 May 1928, pp. 379–435.

19. *Bahr el Shaitan:* for this description I have relied upon the account by Loebsack in *Our Atmosphere*, p. 41.

20. *Inverted mirages in Cuxhaven and Paris:* stories reported by Loebsack in *Our Atmosphere*, pp. 38–44.

21. *Fata morgana:* as described by Fraser and Mach in 'Mirages', p. 102.

22. *Halos, parhelia, sun dogs:* for these effects I have relied upon the information given by Minnaert in *Light and Colour in the Open Air*, pp. 197–202.

23. *Coronae and glories:* see Minnaert, *Light and Colour in the Open Air*, pp. 208–25.

24. *Byrd quotation:* from Admiral Richard E. Byrd, *Alone*, G. P. Putnam's Sons, New York, 1938, as excerpted in *The Armchair Science Reader*, Isabel C. Gordon and Sophie Sorkin, eds., Simon and Schuster, New York, 1959, pp. 462–4.

Chapter 12. Wind from the Sun

1. *Epigraph:* Amy Lowell, 'Night Clouds'.

2. *Sputnik I launch:* the details of the first Soviet launch are based chiefly on William R. Shelton's two books, *Soviet Space Exploration: The First Decade*, Washington Square Press, New York, 1968, pp. 56–63; and *Man's Conquest of Space*, National Geographic Society, Washington, D.C., 1968, pp. 8–10 and 153. Also Lester A. Sobel's *Space: From Sputnik to Gemini*, Facts on File, Inc., New York, 1965, pp. 6–11 and 36.

3. *New York Times quotation:* as quoted by Shelton in *Man's Conquest of Space*, p. 10.

4. *Laika:* the description of Laika, her training and life-support system is based on Shelton's *Man's Conquest of Space*, pp. 79–81 (photograph of Laika); and Sobel's *Space: From Sputnik to Gemini*, p. 8.

5. *Death of Laika:* official reports and statements as cited by Sobel in *Space: From Sputnik to Gemini*, pp. 9–10.

6. *Launch of Vanguard:* as described by Shelton in *Man's Conquest of Space*, p. 12.

7. *Shelton quotation:* from *Man's Conquest of Space*, p. 12.

8. *Launch of Explorer 1:* the description is based on Shelton's account in *Man's Conquest of Space*, pp. 12–17.

9. *Description of Explorer 1 :* based on facts cited by Sobel in *Space: From Sputnik to Gemini*, p. 33; also James A. Van Allen, 'Explorer's News from Space', *Life*, 17 February 1958, p. 51.

10. *Van Allen quotation:* from Van Allen, 'Explorer's News from Space', p. 51.

11. *Altered view of space environment:* as described by J. A. Ratcliffe, *Sun, Earth, and Radio*, McGraw-Hill Book Co., New York, 1970, p. 179.

12. *Solar wind and the magnetosphere:* this description is based principally upon James A. Van Allen, 'Interplanetary Particles and Fields', *Scientific American*, Vol. 233, No. 3, September 1975, pp. 160–73; and Ratcliffe, *Sun, Earth, and Radio*, pp. 180–200.

13. *Temperatures in the ionosphere:* as cited by Chandler in *The Air Around Us*, Natural History Press for the American Museum of Natural History, Garden City, N.Y., 1969, p. 31.

14. *Folklore explanations of aurora:* as related by Loebsack in *Our Atmosphere*, p. 45.

15. *French aurora of 1585:* this story was related by Cook in *Our Astonishing Atmosphere*, Dial Press, New York, 1957, p. 177.

16. *Eyewitness account of auroral lights in Siberia:* Shklovsky, *In Far North-East Siberia*, trans. by L. Edwards and Z. Shklovsky, Macmillan & Co., London, 1916, pp. 105–6.

17. *Auroral lights:* this description is based upon the account given by C. T. Elvey and Franklin E. Roach in 'Aurora and Airglow', *Scientific American*, Vol. 193, No. 3, September 1955, pp. 140–42.

18. *Aurora and sunspot cycles:* the history of the discovery of the relationship between these two phenomena is based upon E. N. Parker's 'The Sun', *Scientific American*, Vol. 233, No. 3, September 1975, pp. 43–50; and John A. Eddy's 'The Maunder Minimum', *Science*, Vol. 192, 18 June 1976, pp. 1189–1202.

19. *Sunspots and solar activity:* the description of these phenomena is based principally upon Parker's 'The Sun', *Scientific American*, Vol. 233, No. 3, September 1975, pp. 42–58.

20. *Evidence of die-off at times of magnetic reversals:* as reported by James D. Hays and Neil D. Opdyke, 'Antarctic Radiolaria, Magnetic Reversals, and Climatic Change', *Science*, Vol. 158, 24 November 1967, pp. 1001–1011.

21. *Temperature measured at 100,000 feet following solar flare:* figure cited by Blumenstock in *The Ocean of Air*, p. 244.

22. *Flares photographed from Skylab:* as reported by Parker in 'The Sun', p. 44.

23. *Apollo solar wind experiment:* as described in *McGraw-Hill Yearbook of Science and Technology: 1970*, McGraw-Hill Book Co., New York, 1971, p. 271.

24. *Parker quotation:* from Parker, 'The Sun', p. 50.

25. *Power-line effect on the ionosphere:* this information came from a personal communication with Robert A. Helliwell, Stanford University.

26. *Earth's radiation belts now 'under human control'*: statement made in *Quality – The Magazine of Product Assurance*, Hitchcock Company, Wheaton, Ill., December 1975, p. 5.

Chapter 13. Rain from the Universe

1. *Epigraph:* Francis Thompson, 'The Mistress of Vision'.

2. *History of cosmic ray research:* the information in this section is based upon Harry Sootin's account in *The Long Search: Man Learns about the Weather*, W. W. Norton & Company, New York, 1967, pp. 189–91.

3. *Millikan and cosmic rays:* details as reported in *The Harper Encyclopedia of Science*, Harper & Row, New York, 1963, p. 294; also Sootin, *The Long Search*, pp. 193–4.

4. *Distribution of cosmic rays:* as reported by V. L. Ginzburg in 'The Astrophysics of Cosmic Rays', *Scientific American*, Vol. 220, No. 2, February 1969, pp. 51–3.

5. *Number of cosmic rays striking human body:* as cited by J. Gordon Cook in *Our Astonishing Atmosphere*, Dial Press, New York, 1957, p. 190.

6. *Primary cosmic rays:* proportion of various atomic nuclei as reported by V. L. Ginzburg, 'The Astrophysics of Cosmic Rays', *Scientific American*, Vol. 220, No. 2, February 1969, p. 52.

7. *Energy of primary rays:* as reported by Ginzburg, 'The Astrophysics of Cosmic Rays', p. 54.

8. *Theories of cosmic ray origin:* based on Ginzburg's account in 'The Astrophysics of Cosmic Rays', p. 63.

9. *Lifetime of cosmic rays:* personal communication from John A. Simpson, University of Chicago.

10. *Mutation rate in human beings:* the figure of two per hundred or 1.6 million defective births per year is taken from Sootin's *The Long Search*, p. 188.

11. *Carbon 14 produced by cosmic rays:* most of my information on this phenomenon was obtained in a personal interview with John A. Simpson of the University of Chicago. I have also relied upon James A. Van Allen's 'Interplanetary Particles and Fields', *Scientific American*, Vol. 233, No. 3, September 1975, pp. 161–73.

12. *Light flashes seen on Skylab mission:* as reported by L. S. Pinsky, W. Z. Osborne, R. A. Hoffman, and J. V. Bailey, 'Light Flashes Observed by Astronauts on Skylab 4', *Science*, Vol. 188, 30 May 1975, pp. 928–30.

13. *Primary ray could cause a flash in an electric light:* statement taken from article by Eugene N. Parker, 'Cosmic Rays', *McGraw-Hill Encyclopedia of Science and Technology*, McGraw-Hill Book Co., New York, 1957.

14. *Helmets worn by astronauts:* as mentioned by I. R. Cameron, 'Meteorites and Cosmic Radiation', *Scientific American*, Vol. 229, No. 1, July 1973, p. 71.

15. *Meteoroids striking spacecraft:* as reported by Shelton in *Man's Conquest of Space*, pp. 68–71.

16. *Zodiacal light:* the description of this phenomenon is based on Minnaert, *Light and Colour in the Open Air*, pp. 290–95.

17. *Space probe measurements of zodiacal light:* my information on this subject came from personal interviews with scientists at N A S A's Ames Research Center; also *McGraw-Hill Yearbook of Science and Technology: 1973*, McGraw-Hill Book Co., New York, 1974, pp. 439–40.

18. *Cooper quotation:* as quoted by Shelton in *Man's Conquest of Space*, p. 118.

19. *Amount of meteor dust entering earth's atmosphere:* figures as cited by Cook in *Our Astonishing Atmosphere*, p. 146.

20. *Bowen's research:* as reported by Chandler in *The Air Around Us*, p. 81; also Loebsack in *Our Atmosphere*, p. 80.

21. *Norton County, Kansas, meteorite:* details as reported in *Science Illustrated*, Vol. 3, November 1948, pp. 22–3.

22. *Siberian meteoroid:* the story of this event, the information that has been collected about it, and the various theories concerning its nature are given in John Baxter and Thomas Atkins, *The Fire Came By*, Doubleday & Company, Garden City, N.Y., 1976.

Chapter 14. Atmospheres of the Solar System

1. *Epigraph:* Arthur O'Shaughnessy, 'Ode'.

2. *Viking 1 landing:* my description of the landing is based upon the account given by G. A. Soffen and C. W. Snyder in 'The First Viking Mission to Mars', *Science*, Vol. 188, 27 August 1976, pp. 759–65.

3. *Viking experiments:* my information on the Viking life experiments and the temperature and atmospheric pressure on Mars is taken from 'Mars Soil Data Still Puzzle Scientists', *Chemical and Engineering News*, 30 August 1976, pp. 23–5.

4. *Mars:* my information on the atmosphere and general conditions of Mars is based on James B. Pollack's 'Mars', *Scientific American*, Vol. 233, No. 3, September 1975, pp. 106–17; and on a personal interview with Pollack in March 1976.

5. *Theory of increased solar radiation in planetary history:* as described by

William K. Hartmann, 'If Mars Once Had Rivers, Was the Sun Then Hotter?', *Smithsonian*, September 1975, pp. 50–55.

6. *McCrea's theory of the galactic year:* W. H. McCrea, 'Ice Ages and the Galaxy', *Nature*, Vol. 255, 19 June 1975, pp. 607–9.

7. *Venus:* my principal sources of information about Venus were the articles by Andrew and Louise Young, 'Venus', *Scientific American*, Vol. 233, No. 3, September 1975, pp. 70–82; and Carl Sagan, 'The Solar System', *Scientific American*, Vol. 233, No. 3, September 1975, pp. 22–32; also personal communications from James B. Pollack and Lawrence Colin at NASA's Ames Research Center, and the series of articles on Pioneer Venus results, *Science*, Vol. 205, No. 4401, 6 July 1979.

8. *Mercury:* my description is based principally on the account given by Bruce C. Murray in 'Mercury', *Scientific American*, Vol. 233, No. 3, September 1975, pp. 58–70.

9. *Planetary magnetic fields:* as discussed by Carl Sagan in 'The Solar System', *Scientific American*, Vol. 233, No. 3, September 1975, p. 29.

10. *Jupiter:* my sources of information about Jupiter were John H. Wolfe, 'Jupiter', *Scientific American*, Vol. 233, No. 3, September 1975, pp. 118–26, and Sagan, 'The Solar System'.

11. *Cause of coloration in Jovian atmosphere:* possibilities proposed by Lewis and Sagan as reported by Allen L. Hammond in 'Exploring the Solar System (V): Atmosphere and Climates', *Science*, Vol. 187, 24 January 1975, p. 244.

12. *Saturn:* the principal sources of my information on Saturn were the articles by Sagan, 'The Solar System', and by Donald M. Hunten, 'The Outer Planets', *Scientific American*, Vol. 233, No. 3, September 1975, pp. 130–42.

13. *Size of ice crystals in Saturn's rings:* personal communication from James B. Pollack.

14. *Uranus and Neptune:* my sources of information on these planets were Hunten, 'The Outer Planets', and Sagan, 'The Solar System'.

15. *Titan:* as described by Gregory Benford in 'Atmospheric Titans', *Natural History*, Vol. 83, No. 4, April 1974, pp. 69–71; also Sagan in 'The Solar System'.

16. *Sagan quotation:* from Sagan, 'The Solar System', p. 24.

17. *Age of the Universe:* history of changing theories as described by Robert Jastrow in 'How Old Is the Universe?', *Natural History*, Vol. 83, No. 7, August–September 1974, pp. 80–82.

18. *Creation of the universe:* in this section I have followed closely the theory presented by A. G. W. Cameron in 'The Origin and Evolution of the Solar System', *Scientific American*, Vol. 233, No. 3, September 1975, pp. 32–41.

19. *Molecules in interstellar spaces:* as reported by Richard McCray in 'Molecules Between the Stars', *Natural History*, Vol. 83, No. 10, December 1974,

pp. 72–7; also personal interview with David Black at N A S A's Ames Research Center.

20. *Formation of the planets:* in this section I have followed the theories of formation as described by Hunten in 'The Outer Planets' and Cameron in 'The Origin and Evolution of the Solar System'.

21. *Formation of the earth:* this description is based principally on the account given by Pollack in 'Mars', p. 110, and on a personal communication from Pollack.

22. *Volcanic cone on Mars:* as described by Pollack, 'Mars', p. 111.

23. *Krafft quotation:* from Maurice and Katia Krafft, *Volcano*, Harry N. Abrams, New York, 1975.

24. *Titov quotation:* as quoted by Shelton, *Man's Conquest of Space*, National Geographic Society, Washington, D.C., 1968, pp. 118–19.

25. *Jeans quotation:* from James Jeans, *The Mysterious Universe*, Cambridge University Press, New York, 1930.

Chapter 15. Fire or Ice

1. *Epigraph:* Marcus Aurelius (Antoninus), *Meditations*, IV, translated by Morris Hickey Morgan.

2. *Raymer's description:* from 'The Nightmare of Famine', *National Geographic*, July 1975, pp. 33–9.

3. *Markandaya quotation:* from Kamala Markandaya, *Nectar in a Sieve*, John Day Co., New York, 1954, p. 120.

4. *Mullen quotation and description:* from William Mullen, 'The Faces of Hunger', *Chicago Tribune*, Part I, 13–18 October, 1974; Part II, 18–21 April 1976.

5. *Erratic weather from 1968 through 1975:* my description is based largely upon Stephen H. Schneider's *The Genesis Strategy*, Plenum Press, New York, 1976, pp. 6–7 and 85–90; also William Mullen's 'Faces of Hunger', Part II.

6. *Droughts of 1976:* as reported in the *Chicago Tribune*, 12 July and 20 August 1976.

7. *Unusually favorable climate between 1955 and 1970:* as reported by Schneider in *The Genesis Strategy*, p. 6, and in a personal interview, March 1976.

8. *Core analysis:* as described by Reid A. Bryson in a personal interview, October 1975.

9. *Results of core analysis:* as reported by Reid A. Bryson in a personal interview and in 'A Perspective on Climatic Change', *Science*, Vol. 184, 17 May 1974, pp. 753–60.

10. *Milankovitch theory:* as reported by J. D. Hays, John Imbrie, and N. J. Shackleton in 'Variations in the Earth's Orbit: Pacemaker of the Ice Ages', *Science*, Vol. 194, 10 December 1976, pp. 1121–32.

11. *Cause of the Great Flood:* as postulated by Cesare Emiliani *et al.*, 'Paleo-climatological Analysis of Late Quaternary Cores from the Northeastern Gulf of Mexico', *Science*, Vol. 189, 26 September 1975, pp. 1083–7.

12. *Emiliani quotation:* from 'Paleoclimatological Analysis of Late Quaternary Cores from the Northeastern Gulf of Mexico', p. 1087.

13. *History of weather in Europe and America from the year 400 to the present:* in this section I have followed the account given by Schneider in *The Genesis Strategy*, pp. 69–78. Analyses by other weather experts may differ slightly in the dates and conditions cited.

14. *Ladurie's study of Alpine glaciers:* as reported by Henry Lansford in 'Climate Outlook: Variable and Possibly Cooler', *Smithsonian*, November 1975, p. 144.

15. *Midwest temperatures in 1830s:* the statement concerning these temperatures was made in the 'Report of the Ad Hoc Panel on the Present Interglacial', Federal Council for Science and Technology, Science and Technology Policy Office, National Science Foundation, August 1974, p. 20.

16. *Northern migration of fish and birds:* as described by Rachel Carson in *The Sea Around Us*, Oxford University Press, New York, 1951, pp. 183–6.

17. *Hubert Lamb's research:* as cited by Lansford in 'Climate Outlook: Variable and Possibly Cooler', *Smithsonian*, November 1975, p. 142.

18. *Temperature changes in the North Atlantic:* as cited by Roger G. Barry, John T. Andrews, and Mary A. Mahaffy, 'Continental Ice Sheets: Conditions for Growth', *Science*, Vol. 190, 5 December 1975, p. 980.

19. *Little Ice Age and solar activity:* for a discussion of the possible relationship, see, for example, Stephen H. Schneider and Clifford Mass, 'Volcanic Dust, Sunspots, and Temperature Trends', *Science*, Vol. 190, 21 November 1975, pp. 741–2.

20. *Sunspot activity and wheat crops:* possible correlation cited by John Gribben in 'Whither the Weather', *Natural History*, Vol. 84, No. 5, June–July 1975, pp. 8–17.

21. *Droughts and sunspots:* possible correlation mentioned by Stephen H. Schneider in personal communication.

22. *Weather and the magnetic cycle of the solar wind:* my information on this subject came from a personal interview with John M. Wilcox; also John M. Wilcox, 'Solar Structure and Terrestrial Weather', *Science*, Vol. 192, 21 May 1976, pp. 745–8.

23. *Volcanic eruptions and colder weather:* possible correlation as discussed by Schneider and Mass in 'Volcanic Dust, Sunspots, and Temperature Trends'. Evidence of poor growing seasons and freezes following volcanic eruptions was found by scientists at the University of Arizona's tree-ring laboratory, as reported by *Time*, 24 March 1975, p. 64.

24. *Volcanic activity and ice ages:* evidence of correlation from deep-sea cores cited by James B. Pollack *et al.*, 'Stratospheric Aerosols and Climatic Change', submitted to *Nature*.

25. *Change in size of earth's polar caps:* as reported by Gribben in 'Whither the Weather', p. 14.

26. *The deserts are growing:* figures given by Erik P. Eckholm of Worldwatch Institute in an address to the American Association for the Advancement of Science, citing the United Nations study as reported by the *New York Times*, 7 March 1976.

27. *Loss of cultivable acres to the Thar Desert:* figure cited by Schneider, *The Genesis Strategy*, p. 83.

28. *Decrease in sunlight in U.S.A.:* figures compiled by N O A A, reported by the *Wall Street Journal*, 20 March 1975.

29. *Bryson's theory of the dust veil and climatic change:* as set forth in Reid A. Bryson, 'A Perspective on Climatic Change', *Science*, Vol. 188, 13 June 1975, pp. 753–60.

30. *Estimates of effects of increasing carbon dioxide on the earth's climate:* figures as cited by Schneider in *The Genesis Strategy*, p. 135; also by Bolin in 'The Carbon Cycle', *Scientific American*, Vol. 223, No. 3, September 1970, p. 131.

31. *Fluorocarbons enhance greenhouse effect:* my discussion of this effect is based on *Halocarbons: Environmental Effects of Chlorofluoromethane Release*, pp. 6–8.

32. *Waste heat from energy production:* the possibility that exponential increase in energy usage could cause serious climatic change is expounded by Howard A. Wilcox in *Hothouse Earth*, Praeger Publishers, New York, 1975. It is also discussed by Schneider in *The Genesis Strategy*, pp. 166–71 and Appendix B.

33. *Sagan quotation:* from Sagan, 'The Solar System', p. 27.

34. *World's grain reserve:* as cited by Schneider in *The Genesis Strategy*, pp. 99–103; and Lester R. Brown in *By Bread Alone*, Praeger Publishers, New York, 1974, pp. 58–62. See also Lester R. Brown, 'Increasingly the U.S. Is Breadbasket to the World', the *New York Times*, 7 December 1975.

35. *C I A statement:* from U.S. Central Intelligence Agency, *Potential Implications of Trends in World Population, Food Production, and Climate*, O P R-401, Washington, D.C., August 1974, pp. 39–41.

36. *Rumored weather offensive against Cuba:* as reported by United Press

International and the *New York Times*, 28 June 1976. The information was released by Lowell Ponte, former Defense Department consultant.

37. *Plan to spread soot on the glaciers:* described by Schneider in *The Genesis Strategy*, pp. 206–7.

Chapter 16. The Ninety-Year Forecast

1. *Epigraph:* Edna St Vincent Millay, 'God's World'.

2. *Thomas quotation:* from Lewis Thomas, *The Lives of a Cell*, Viking Press, New York, 1974, p. 145.

3. *Woolf quotation:* as quoted by David Cecil in 'Virginia Woolf', *Poets and Story-Tellers*, Macmillan Company, New York, 1949, p. 161.

4. *Eiseley quotation:* from Loren Eiseley, *The Firmament of Time*, Atheneum, New York, 1960, p. 57.

Acknowledgements

1. *Sears quotation:* from *Where There Is Life*, Dell Publishing Co., New York, 1962, p. 30.

SUGGESTIONS FOR FURTHER READING

BATTAN, LOUIS J., *Cloud Physics and Cloud Seeding*, Doubleday & Company, Garden City, New York; Educational Services Inc., Somerville, N.J., 1962; Heinemann Education, London, 1965.

BATTAN, LOUIS J., *The Unclean Sky*, Doubleday & Company, Garden City, New York; Educational Services Inc., Somerville, N.J., 1966. These books are simplified accounts of meteorology and pollution written for a lay audience.

BAXTER, JOHN, and ATKINS, THOMAS, *The Fire Came By*, Doubleday & Company, Garden City, New York, 1976; Futura Publications, London, 1977; description and speculation about the great meteor that fell in Siberia in 1908.

BLUMENSTOCK, DAVID I., *The Ocean of Air*, Rutgers Universal Press, New Brunswick, N.J., 1959; Macdonald, London, 1976. A fine general book about the earth's atmosphere.

BROWN, LESTER R., *By Bread Alone*, Praeger Publishers, New York, for the Overseas Development Council, 1974; Pergamon Press, London, 1975. Deals with the problem of world food supply.

BROWN, SLATER, *World of the Wind*, Bobbs-Merrill Company, Indianapolis, 1961. Tells many interesting facts and stories about winds around the world.

BRYSON, REID A., and MURRAY, THOMAS J., *Climates of Hunger*, University of Wisconsin Press, Wisconsin, 1977. Evidence of changing climates over the past three or four thousand years.

CALDER, NIGEL, *The Weather Machine*, Viking Press, New York, 1974; BBC, London, 1974. A fine recent book on climatology for the layman. Contains many excellent photographs and diagrams.

FRANCIS, PETER, *Volcanoes*, Penguin Books, New York and London, 1976. An up-to-date and absorbing study of volcanic activity.

GEORGE, UWE, *In the Deserts of this Earth*, Harcourt Brace Jovanich, New York and London, 1977 (translated from the German by Richard and Clara Winston).

GLINES, C. V., ed., *Lighter-than-Air Flight*, Franklin Watts, 1965. A collection of first-hand or contemporary accounts of historic ascents.

HALACY, D. S., Jr, *The Weather Changers*, Harper & Row, New York, 1968. Description for the layman of climate modification techniques.

KRAFFT, MAURICE and KATIA, *Volcano,* Harry N. Abrams, New York, 1975; stunning photography and interesting text.

LAUSANNE, S. A., ed., *The Romance of Ballooning*, Viking Press, New York, 1971. History and facts about ballooning.

LINCOLN, JOSEPH COLVILLE, ed., *On Quiet Wings*, Northland Press, Flagstaff, Arizona, 1972.

LINCOLN, JOSEPH COLVILLE, ed., *Soaring on the Wind*, Northland Press, Flagstaff, Arizona, 1972. These two books contain beautiful photographs and descriptions of gliding.

MINNAERT, M., *Light and Colour in the Open Air*, translated by H. M. Kremer-Priest; revised by K. E. Brian Jay, G. Bell & Sons, London, 1940. A very complete authoritative description of meteorological optics.

MURCHIE, GUY, *Song of the Sky*, Riverside Press, Cambridge, Massachusetts; Houghton Mifflin Co., Boston, 1954; fact and poetry woven into a delightful book for the general reader.

The Rainbow Book, The Fine Arts Museum of San Francisco in association with Shambhala, Berkeley and London; Fine Arts Museums of San Francisco, San Francisco, 1975.

SCHNEIDER, STEPHEN H., with MESIROW, LYNNE E., *The Genesis Strategy*, Plenum Press, New York, 1976. Packed with interesting information about weather history and theories of long-term weather changes.

SOOTIN, HARRY, *The Long Search: Man learns about the Weather*, W. W. Norton & Company, 1967. A general account of climatology for the layman.

STEHLING, KURT R., and BELLER, WILLIAM, *Skyhooks*, Doubleday & Company, Garden City, N.Y., 1962. Contains many exciting and interesting tales about man's exploration of the atmosphere.

WINKLESS, NELS, III, and BROWNING, IBEN, *Climate and the Affairs of Man*, Harper's Magazine Press, New York, 1975; P. Davies, London, 1976. Presents a theory of climatic change influenced by volcanic activity.

WISE, WILLIAM, *Killer Smog*, Rand McNally & Company, Chicago, 1968. A detailed description of the great London smog of 1952.

ACKNOWLEDGEMENTS

The subject matter of *Earth's Aura* touches on many different fields of science and much of it is on the leading edge of an advancing wave of knowledge. It would not have been possible for me, working alone, to have collected all the material which has gone into the composition of this book. I have been fortunate in having the help and cooperation of many distinguished scientists. In some cases they read chapters dealing with their fields of expertise and made suggestions which I incorporated into the text. For such very constructive help I am indebted to Cheves T. Walling, professor of chemistry, University of Utah; Stephen H. Schneider, National Center for Atmospheric Research; John A. Simpson, professor of physics and director of the Enrico Fermi Institute, University of Chicago; James B. Pollack, National Aeronautics and Space Administration's Ames Research Center; John M. Wilcox, professor of physics, Stanford University; and Edward W. Pearl, meteorologist, Department of Geophysical Sciences, University of Chicago.

Many other scientists have also been generous with their time, answering my questions, describing their fields of special interest, and providing valuable source materials. Among these, I especially want to thank Chester W. Newton and John C. Gille of the National Center for Atmospheric Research; T. Theodore Fujita and Roscoe R. Braham, Jr, professors of geophysical sciences, University of Chicago; Robert A. Helliwell, professor of physics, Stanford University; Richard A. Wolf and Joseph W. Chamberlain, professors of physics at Rice University; William E. Gordon, dean of the School of Natural Sciences, Rice University; Lawrence Colin and John H. Wolfe, National Aeronautics and Space Administration's Ames Research Center; and Reid A. Bryson, professor of meteorology and director of the Institute of Environmental Studies, University of Wisconsin.

I also want to express my appreciation to Lucy Drews, research librarian at the Field Museum of Natural History, who spent many hours running down obscure facts and elusive bits of information in the biological sciences, to Phyllis Dubnick, Virginia Bauman, and Beverly Lawson, who helped me put the manuscript in finished form. And finally I am much indebted to my editor, Charles Elliott, who guided ˙.nd encouraged me throughout the long gestation period. In spite of the competent assistance I have received from many experts I am very much aware of my own obligations in presenting the scientific material fairly and correctly. Final responsibility for the accuracy and validity of the text is mine alone.

Readers who are acquainted with my earlier work may be surprised to see that this book devotes less than a fifth of its space to the problems of deteriorating air quality and increasing levels of pollution. But I hope that this book will make many new friends for the environmental movement by calling attention to the beauty, the complexity, and the sensitivity of the remarkable medium in which we spend all our lives. As Paul Sears says, a treasure is best protected by those who understand its worth.

L.B.Y.
Winnetka, Illinois
November 1976

Grateful acknowledgement is made to the following for permission to reprint previously published material:

Harry N. Abrams, Inc. and Draeger Frères: Excerpts from *Volcano* by Maurice and Katia Krafft, taken from pages 13–21. Copyright © 1975 by Draeger, Imprimeurs, Paris.

Bobbs Merrill Co., Inc.: Excerpts from Will Keller quotation in *The World of the Wind* by Slater Brown, pages 199–200. Copyright © 1961 by Slater Brown.

The Chicago Tribune: Excerpts from 'The Faces of Hunger' by William Mullen, Part I, 13–18 October, 1974, and Part II, 18–21 April, 1976.

Columbia University Press: Excerpts from *The Physical Treatises of Pascal*, I. H. B. and A. G. H. Spiers, translators; F. Barry, editor. Copyright 1937 by Columbia University Press, renewed 1965.

Curtis Brown, Ltd: Excerpts from *Killer Smog* by William Wise. Copyright © 1970 by William Wise.

McGraw-Hill Book Company: Excerpts from *Wild Ocean* by Alan J. Villiers, pages 12–16. Copyright © 1957 by Alan J. Villiers.

Natural History Magazine: Excerpts from *Twilight Seen From Space* by K. Ya Kondratyev and O. I. Smotky, *Natural History* Magazine, November 1973. Copyright © 1973 by The American Museum of Natural History.

Pantheon Books, a division of Random House, Inc., and William Collins Sons & Co., Ltd: Excerpts from *Our Atmosphere* by Theo Loebsack, translated by E. L. and D. Rewald, pages 122 and 124. Copyright © 1959 by William Collins Sons & Co., Ltd, and Pantheon Books, Inc.

G. P. Putnam's Sons: Excerpt from *Alone* by Admiral Richard E. Byrd. Copyright 1938 by Richard E. Byrd; renewed 1966 by Marie A. Byrd.

Justus J. Schifferes: 'Franklin's Letter' from *The Autobiography of Science* (2nd edition), edited by Forest Ray Moulton and Justus J. Schifferes.

Soaring Magazine and Laurence E. Edgar: Excerpts from 'Frightening Experience During Jet-Stream Project' by Laurence E. Edgar. Reprinted from *Soaring*, July–August 1955.

University of California Press: Excerpts from *Mathematical Principles of Natural Philosophy and His System of the World*, by Isaac Newton; translated by Andrew Motte; revised by Florian Cajori. Copyright © 1962 by Florian Cajori.

The Viking Press: Excerpts from *The Lives of a Cell*, by Lewis Thomas. Copyright © 1973 by the Massachusetts Medical Society.

INDEX